Geographical Skills
for Edexcel GCSE in Geography A

UNIT 1

Steph Warren

Orders: please contact Bookpoint Ltd, 130 Milton Park, Abingdon, Oxon OX14 4SB. Telephone: (44) 01235 827720. Fax: (44) 01235 400454. Lines are open 9.00–5.00, Monday to Saturday, with a 24-hour message answering service. Visit our website at www.hoddereducation.co.uk

First published in 2010 by
Hodder Education,
An Hachette UK Company
338 Euston Road
London NW1 3BH

Impression number 5 4 3 2
Year 2014 2013 2012 2011

Illustrations by Gray Publishing and David Gardner
Typeset in 11/13pt Myriad and produced by Gray Publishing, Tunbridge Wells
Printed and bound in Italy

A catalogue record for this title is available from the British Library

ISBN: 978 1444 112344

Contents

Unit 1 Geographical Skills and Challenges

Section A Geographical Skills

Acknowledgements

The Publishers would like to thank the following for permission to reproduce copyright material:

Photo credits
p.1 *t, b* © Steph Warren; **p.6** *t* © Skyscan/CLI; **p.6** *b* © Steph Warren; **p.7** © Steph Warren; **p.9** *t* © Steph Warren; **p.9** *b* © Skyscan.co.uk; **p.10** *t* Image courtesy of Earth Sciences and Image Analysis Laboratory, NASA Johnson Space Center (ISS012-E-15551 http://eol.jsc.nasa.gov); **p.10** *b* © Skyscan.co.uk; **p.11** © Steph Warren; **p.22** © Steph Warren; **p.30** © Steph Warren; **p.31** © English Heritage. NMR. Aerofilms Collection; **p.33** *t, b* © GeoPerspectives supplied by Skyscan.co.uk; **p.35** © Steph Warren; **p.36** © Steph Warren; **p.38** © Steph Warren; **p.39** © Steph Warren; **p.41** © Steph Warren; **p.79** *tl, tr, m, bl, bl* © Steph Warren; **p.86** With permission of Peak Electronic Design Limited; **p.89** *tl* Robert Gray; **p.89** *b* © Steph Warren.

Maps on **pp.42–46**, reproduced from Ordnance Survey mapping with permission of the Controller of HMSO.

1 Basic Skills

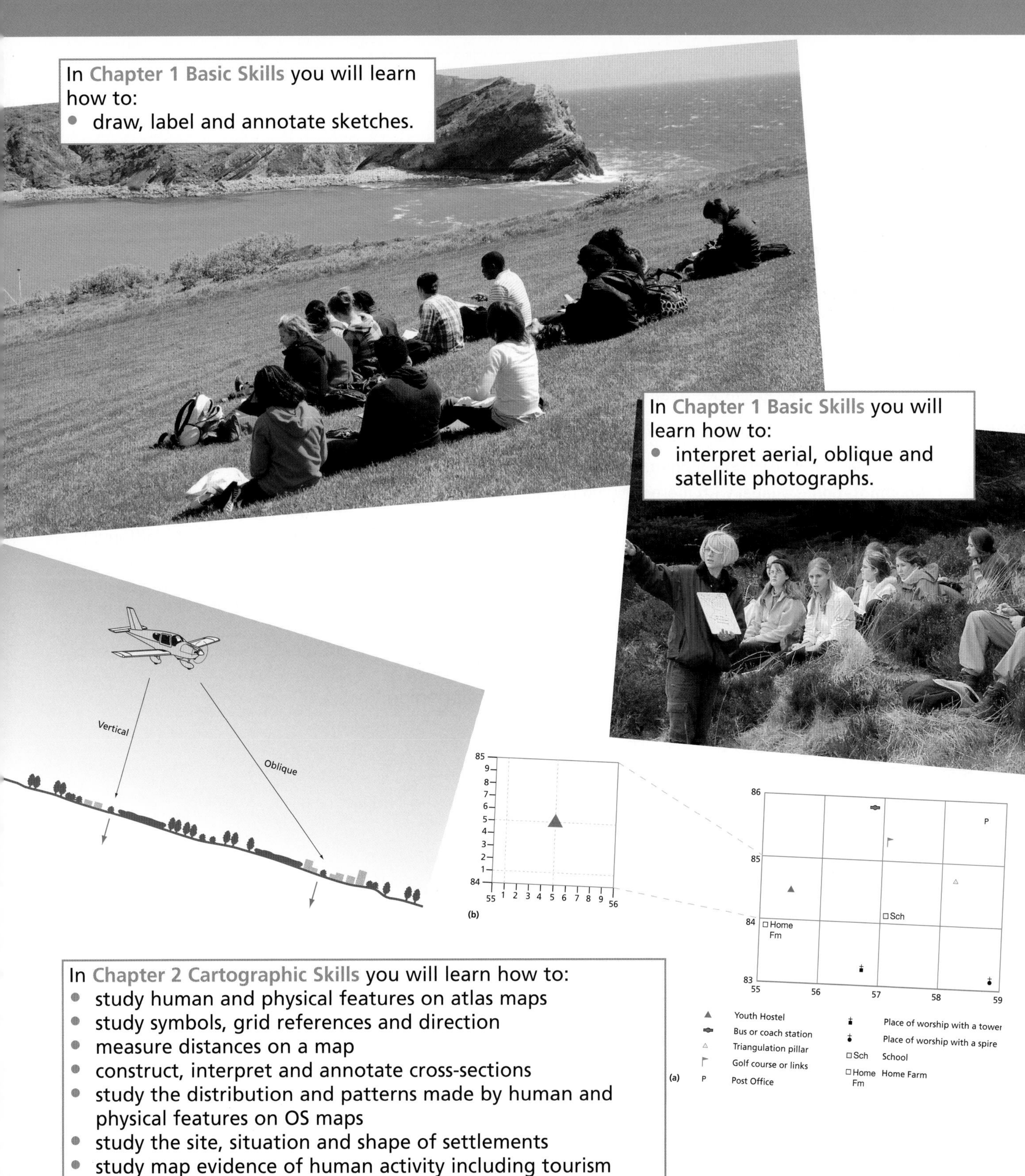

In **Chapter 1 Basic Skills** you will learn how to:

- draw, label and annotate sketches.

In **Chapter 1 Basic Skills** you will learn how to:

- interpret aerial, oblique and satellite photographs.

In **Chapter 2 Cartographic Skills** you will learn how to:

- study human and physical features on atlas maps
- study symbols, grid references and direction
- measure distances on a map
- construct, interpret and annotate cross-sections
- study the distribution and patterns made by human and physical features on OS maps
- study the site, situation and shape of settlements
- study map evidence of human activity including tourism
- study the use of maps in association with photographs, sketches and directions.

Diagrams, graphs and sketch maps

Learning objective – to learn how to draw, label and annotate sketches.

Learning outcomes
- To be able to draw, label and annotate diagrams and graphs.
- To be able to draw, label and annotate sketch maps.

How to label and annotate diagrams, graphs and sketch maps

All geography students need the basic skills of being able to draw, label and annotate diagrams, graphs and sketch maps. These skills will be required to achieve good grades in all geography exams.

What is the difference between a label and an annotation?
- A **label** is a simple descriptive point.
- An **annotation** is a label with more detailed description or an explanatory point.

Exam Tips
- It is a good idea to draw diagrams when explaining the formation of landforms.
- It is a good idea to put a hand-drawn sketch map in your controlled assessment study.

Diagrams

Diagrams should be used in geography to demonstrate knowledge. They can be particularly useful where you are required to explain the formation of landforms. For example, the diagram in Figure 1 shows how cliffs retreat. It can sometimes be more appropriate to draw a series of diagrams to show the sequence of something happening.

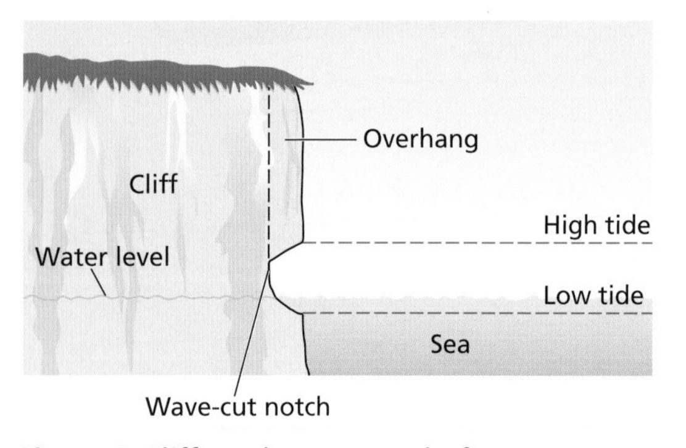

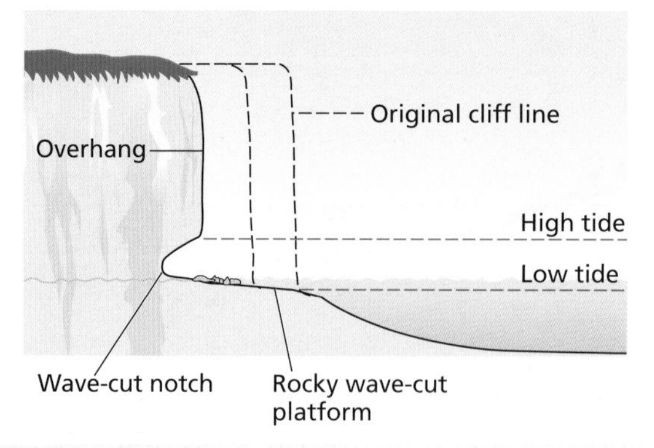

Figure 1 Cliffs and wave-cut platforms

Graphs

Graphs can be labelled to point out changes and patterns, and annotated to explain what is happening. How to draw different graphs is dealt with in Chapter 3, Graphical Skills, page 47. The population pyramid in Figure 2 has been labelled on the left-hand side to point out the important features. The right-hand side contains annotations which develop the points a little further and therefore are not simple labels.

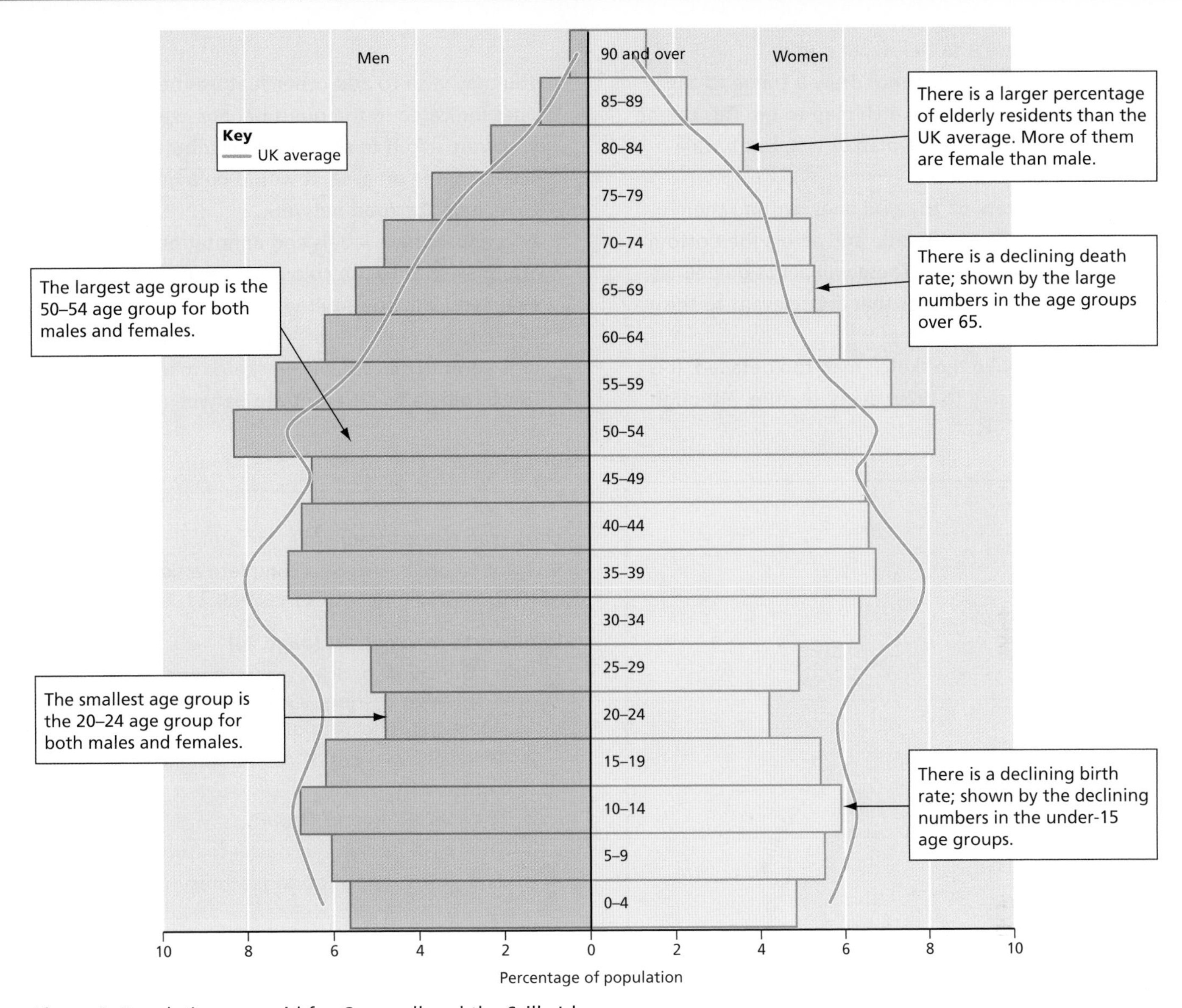

Figure 2 Population pyramid for Cornwall and the Scilly Isles

Sketch maps

Many students are daunted by the idea of drawing sketch maps. What must be remembered is that a perfect replica of the map is not being expected, just a simple representation of the main features. There is an example of a sketch map of part of the Looe OS map (from page 45) on page 4.

During an exam, you may be asked to draw a sketch map of a particular feature, for example, the woodland or the relief and drainage of an area. You must not redraw the whole of the map extract. What you must do is to show all of the features you have been asked for, plus one or two other distinctive features. If you are asked to draw the relief of an area you could simply shade in the area of land above a certain height. The sketch should be clear and well labelled or annotated depending on the command word in the question.

How to draw a sketch from a map

- Using a ruler and pencil draw a frame to the size you want the sketch map to be. This can be the same size as the map or not, but a scale must be included.
- Lightly draw on the grid lines, writing their numbers down the side and across the bottom of the sketch map. These will act as guidelines.
- Draw in the features that are relevant to the question.
- Do not make the sketch map too detailed. It is not necessary to draw **every** feature, although you may wish to add other features besides the ones indicated in the question. For example, if you were asked to sketch the distribution of settlement in an area, it would be a good idea to include the road network.
- Add appropriate labels and annotations.
- You may wish to use colour. However, in the exam you will be required to only use a pencil/black pen. Therefore, it would be a good idea to practise completing sketch maps without using colours to differentiate between features.

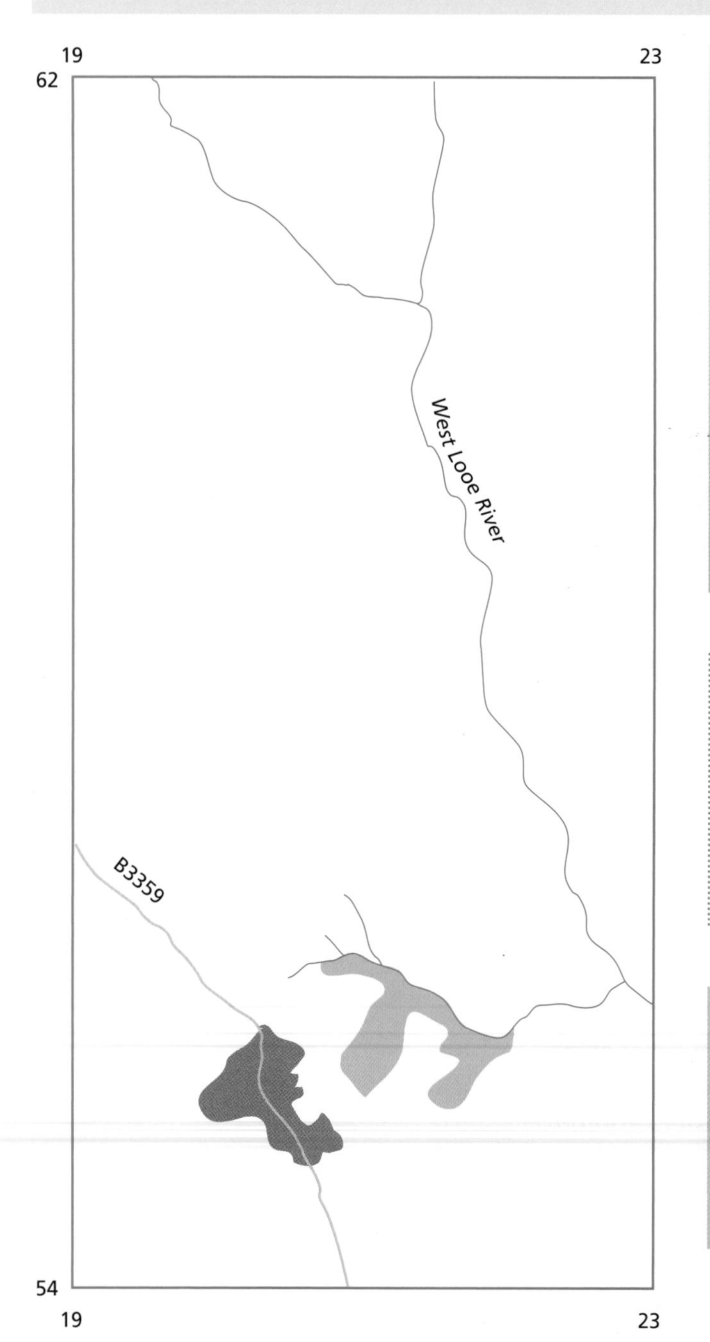

Figure 3 Incomplete sketch map of part of the Looe OS map extract

ACTIVITIES

Looe map extract (page 45)

1 Trace Figure 3. Use it to complete a sketch map of the woodland west of easting 23.

Warkworth map extract (page 44)

2 Using the OS map, draw a sketch map of the area between eastings 23 to 27, northings 05 to 12. Mark the following on your sketch:
   a The built-up areas of Alnmouth and Warkworth.
   b The Rivers Aln and Coquet.
   c The coastline and all coastal features.
   d Any tourist information features in the area.

STRETCH AND CHALLENGE

Using the OS map for Cambridge on page 42, draw a sketch map for the area between eastings 46 to 51 and northings 64 to 67. Include all the human and physical features you feel are relevant to show what the area is like.

Review

By the end of this section you should be able to:

- draw, label and annotate diagrams and graphs
- draw, label and annotate sketch maps.

Drawing and labelling sketches

Learning objective – to learn how to draw, label and annotate sketches from photographs and in the field.

Learning outcomes
- To be able to draw sketches from photographs and in the field.
- To be able to label and annotate sketches.

In the last section you learnt the difference between a label and an annotation. In this section you will learn how to apply your knowledge to sketches. To draw a geographical sketch you do not have to be an excellent artist, you just need to follow some basic steps.

How to draw a sketch from a photograph
- Draw a frame to the size you want the sketch to be.
- Lightly draw lines dividing the frame into four quarters. These will help you to draw the rest of the sketch, acting as guidelines. The lines can be erased when the sketch is completed.
- Draw in the most important lines, such as rivers, coastline and the outline of the hills.
- Draw in the less important features, like woodland, settlements and communication lines. Do not make the sketch too detailed. It is not necessary to draw every feature.
- Add appropriate labels and annotations.
- Rub out the lightly drawn lines that divided the sketch to start off.

The sketch in Figure 4 is a drawing of Warkworth taken from the aerial photograph in Figure 5. It has been given descriptive labels (in blue) and annotations (in red) explaining its site.

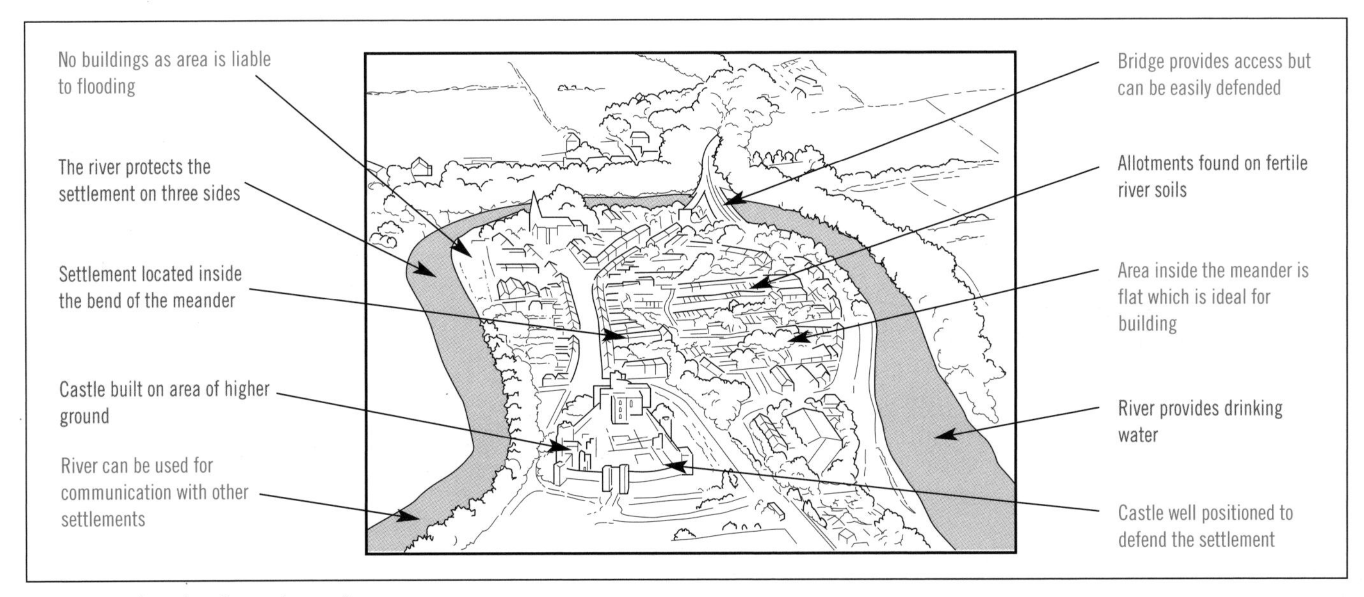

Figure 4 Sketch of Warkworth

Figure 5 Aerial photograph of Warkworth

Figure 6 Photograph of Warkworth from Amble

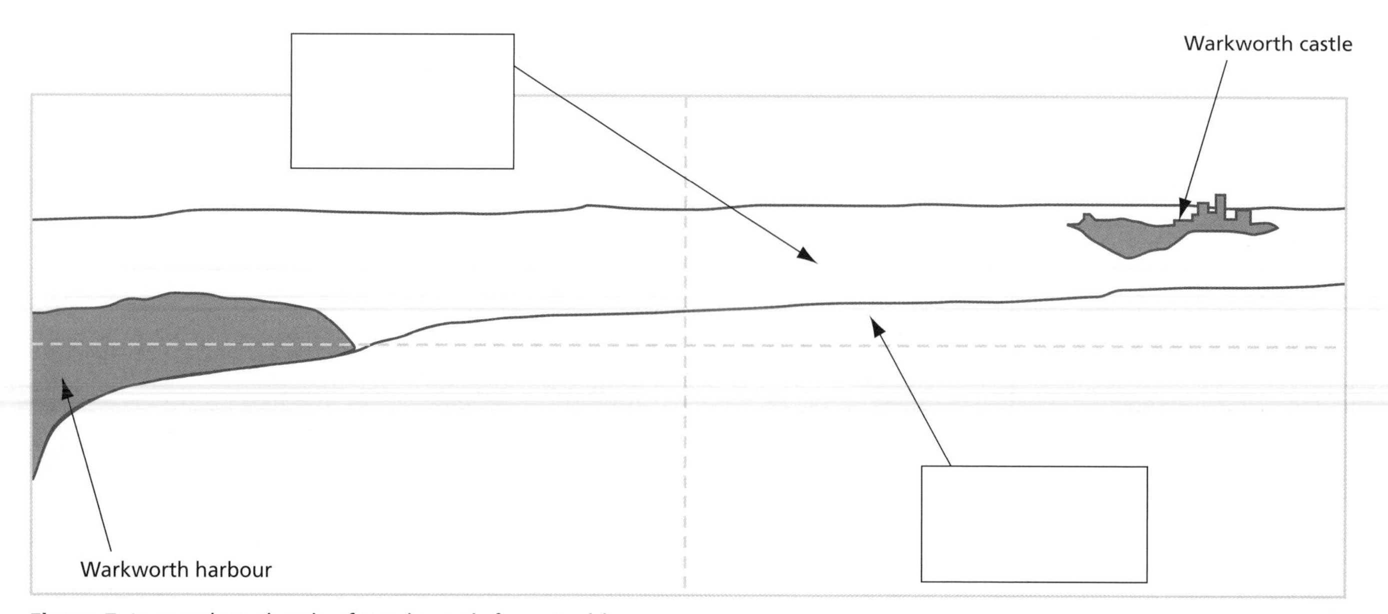

Figure 7 Incomplete sketch of Warkworth from Amble

Figure 8 Photograph of St Bees in Cumbria

ACTIVITIES

1 Study Figure 6. It is a photograph of Warkworth from Amble. Trace Figure 7 – an incomplete sketch of the photograph. Complete the sketch as directed in the 'how to' box on page 5. The boxes give ideas of where you could add labels.

2 Draw a sketch of Figure 8. Label the following on the sketch:
   a A caravan site.
   b Types of coastal defence.
   c The beach.
   d The cliffs.
   e The village of St Bees.
   f The Lake District hills in the background.

Exam Tips

- It is a good idea to include at least one field sketch in your controlled assessment study.
- In exams, you will be asked to complete sketches rather than draw them from scratch.

STRETCH AND CHALLENGE

On your sketch of Warkworth (from question 1 of the Activities), add some annotations to explain the site of Warkworth.

Review

By the end of this section you should be able to:

- draw sketches from photographs and in the field
- label and annotate sketches.

Aerial, oblique and satellite photographs

Learning objective – to learn how to interpret aerial, oblique and satellite photographs.

Learning outcomes
- To be able to interpret aerial and oblique photographs.
- To be able to interpret satellite photographs.

How to interpret aerial, oblique and satellite photographs

Aerial photographs can be taken vertical (directly above). If so they are called 'bird's eye view' photographs because they show a picture taken as if looking down like a flying bird. Other aerial photographs are oblique (taken from an angle) and therefore show more detail such as the sides of buildings. Figure 9 below shows the difference between aerial and oblique photographs.

A satellite image is a picture of the Earth taken from space. Satellite images can show patterns on a large scale, such as the lights from urban areas on a continent, or can zoom in to see small details such as cars on a street.

All of these different types of photographs show elements of the landscape which are not found on OS maps, such as types of crops being grown and the different uses of buildings.

Ensure that you know the difference between vertical and oblique aerial photographs.

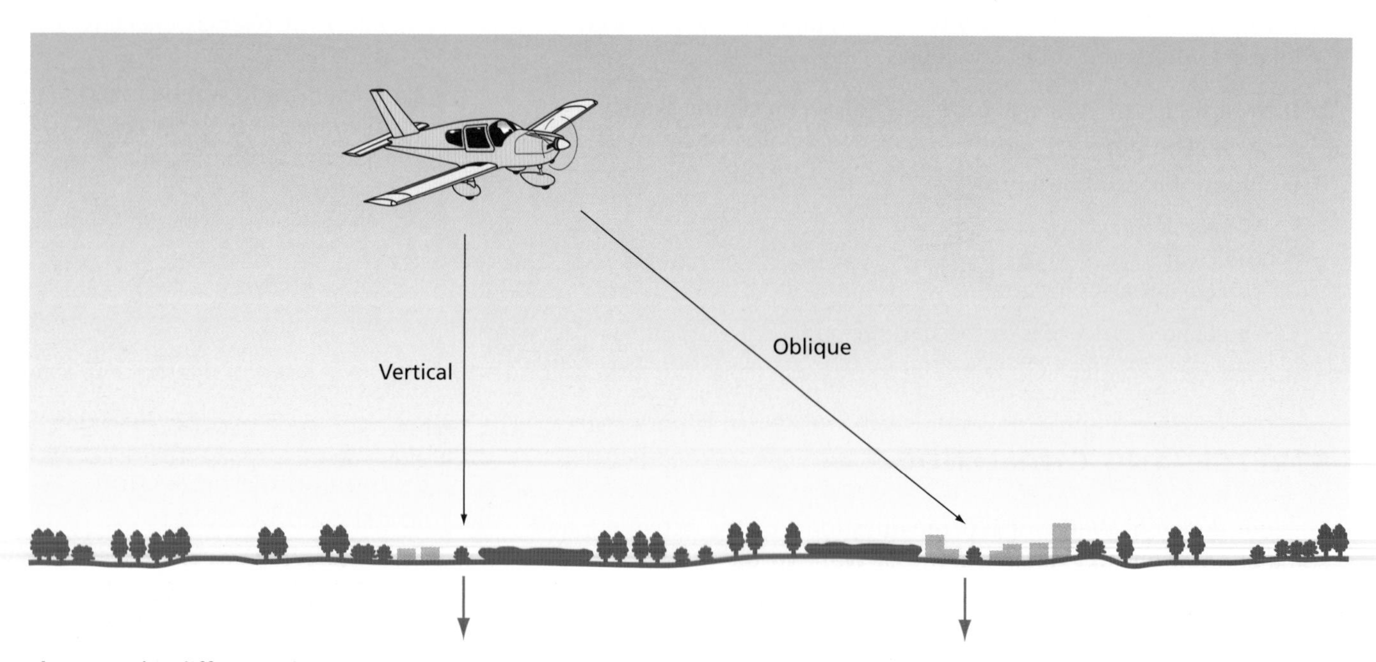

Figure 9 The difference between aerial and oblique photographs

Figure 10 Oblique aerial photograph of Swanage

Figure 11 Vertical aerial photograph of Swanage

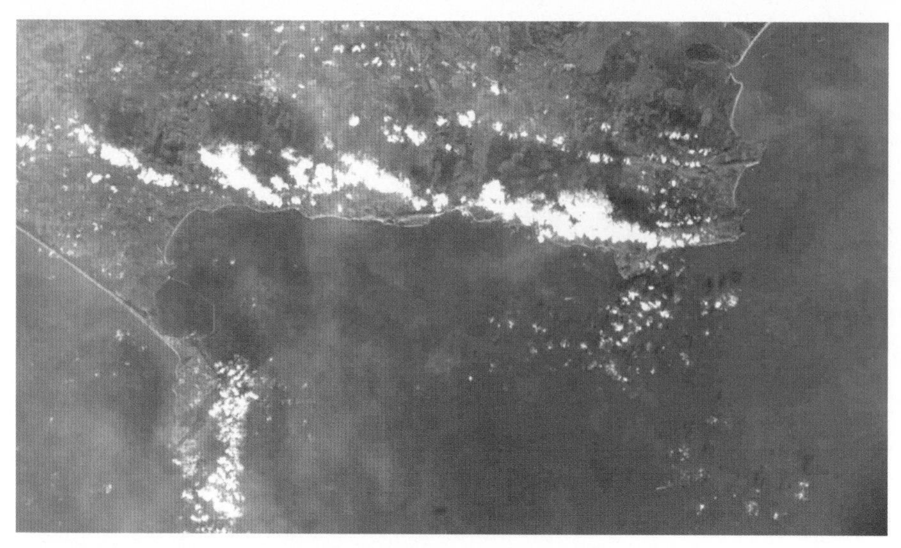

Figure 12 A satellite photograph of the Swanage area

Interpretation of aerial photography

When a photograph is interpreted, it involves describing and explaining the physical and human geography which can be seen on the photograph. It is important when interpreting photographs that the writing is coherent and shows good literacy skills in expressing geographical points.

Example of an interpretation of an aerial photograph

Physical features – relief
The land is very flat with no visible slopes. There are no rivers visible on the photograph. The soil is fertile due to the wide variety of crops being grown.

Human features – land use
Farmland: The fields are large if they are compared with the individual farm buildings shown on the photograph. They are mostly square and rectangular. This implies the use of large arable machinery. It is arable land with a wide variety of crops being grown; this can be seen by the different colours. There are also lines in the fields which show that machinery has been used. Because crops are growing in most of the fields, the photograph must have been taken in the spring or summer. Most of the fields are divided by roads and ditches with the only exception being tree boundaries in the north-east of the photograph which might be being used as a wind break to protect the village of Landbeach.

Settlement: The village of Landbeach, which is clearly shown to the east of the photograph, is a linear settlement. For the length of the village there is only one house on each side of the road (the road number can be added by reference to the OS map). To the west on the map there are farms which display a dispersed land-use pattern.

Figure 13 An aerial photograph of Landbeach near Cambridge and its surroundings

ACTIVITIES

1 Study the different photographs of Swanage shown in Figures 10, 11 and 12. What do the different photographs show you? It may help to put your answer in a table like the one below:

Oblique aerial photograph	Vertical aerial photograph	Satellite photograph

2 Study the aerial photograph of Warkworth on page 6. Describe the human and physical features of the area.

Figure 14 Oblique aerial photograph of Gatwick Airport

- Photographs are used in exams to help you to identify features.
- You may also be asked to draw a sketch from a photograph.

STRETCH AND CHALLENGE

Study Figure 14 above, an aerial photograph of Gatwick Airport. Draw an annotated sketch of the photograph.

Review

By the end of this section you should be able to:

- interpret aerial and oblique photographs
- interpret satellite photographs.

Atlas maps

Learning objective – to study human and physical features on atlas maps.

Learning outcomes
- To recognise and describe distributions and patterns of human features.
- To recognise and describe distributions and patterns of physical features.

Atlases not only contain maps showing where places are, but also show physical patterns such as the height of the land and human patterns such as population density. In an exam, you will need to be able to describe patterns of human geography and patterns of physical geography and relate them to each other. Atlases contain maps at a variety of different scales, but the most common ones show patterns on a country, continent and world scale. You should be familiar with all of these scales.

How to describe the pattern of physical or human features on an atlas map
- Begin with a general statement about where the features are located on the map. For example on Figure 1a, China is densely populated on the eastern side of the country.
- Then go into greater detail such as mentioning the area of the country or any particular features such as the name of the sea next to that area, in this case the Yellow Sea.
- You should then be more specific and include in the answer where the population is concentrated, such as the River Yangtze.

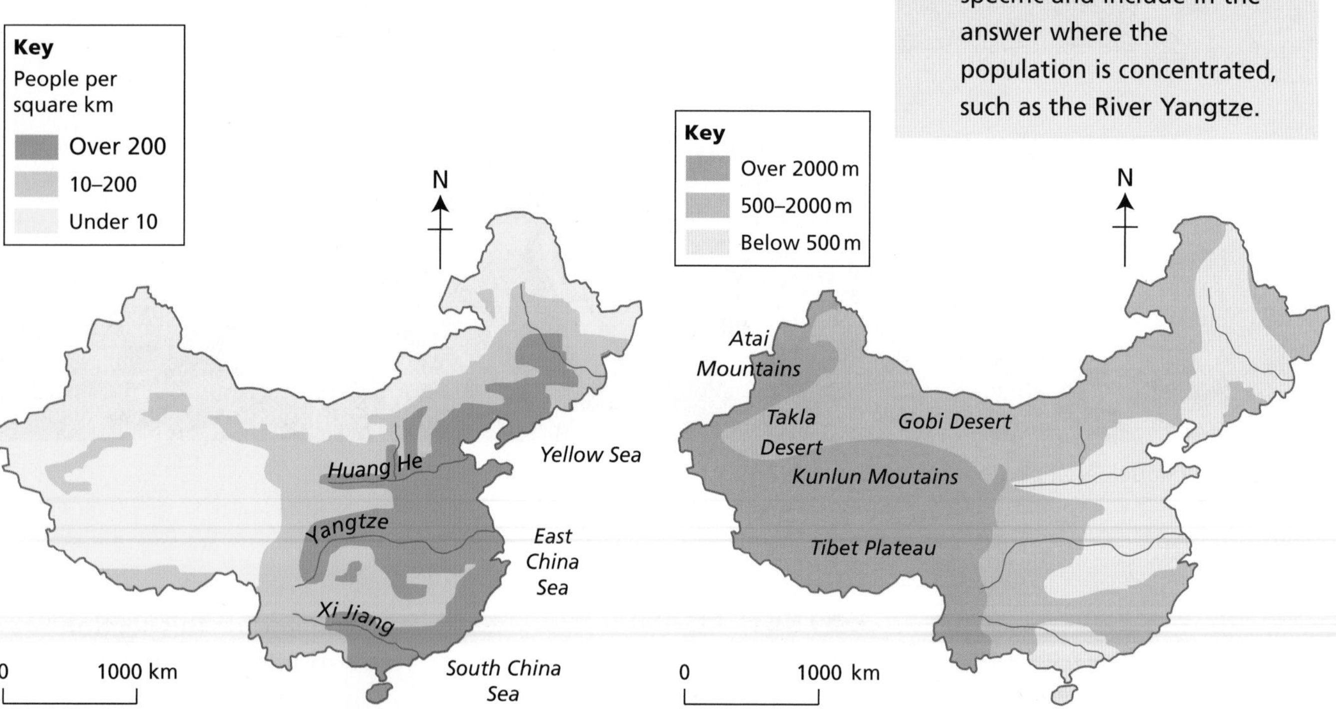

Figure 1a Population density of China

Figure 1b Physical features of China

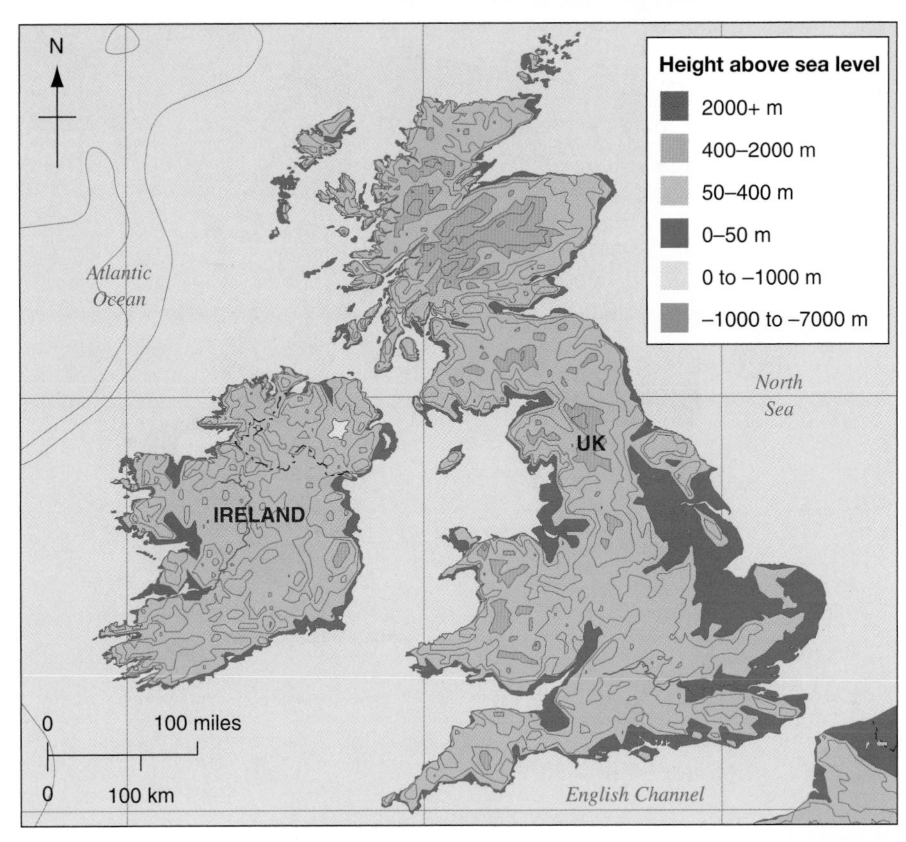

Figure 2 A relief map of the British Isles

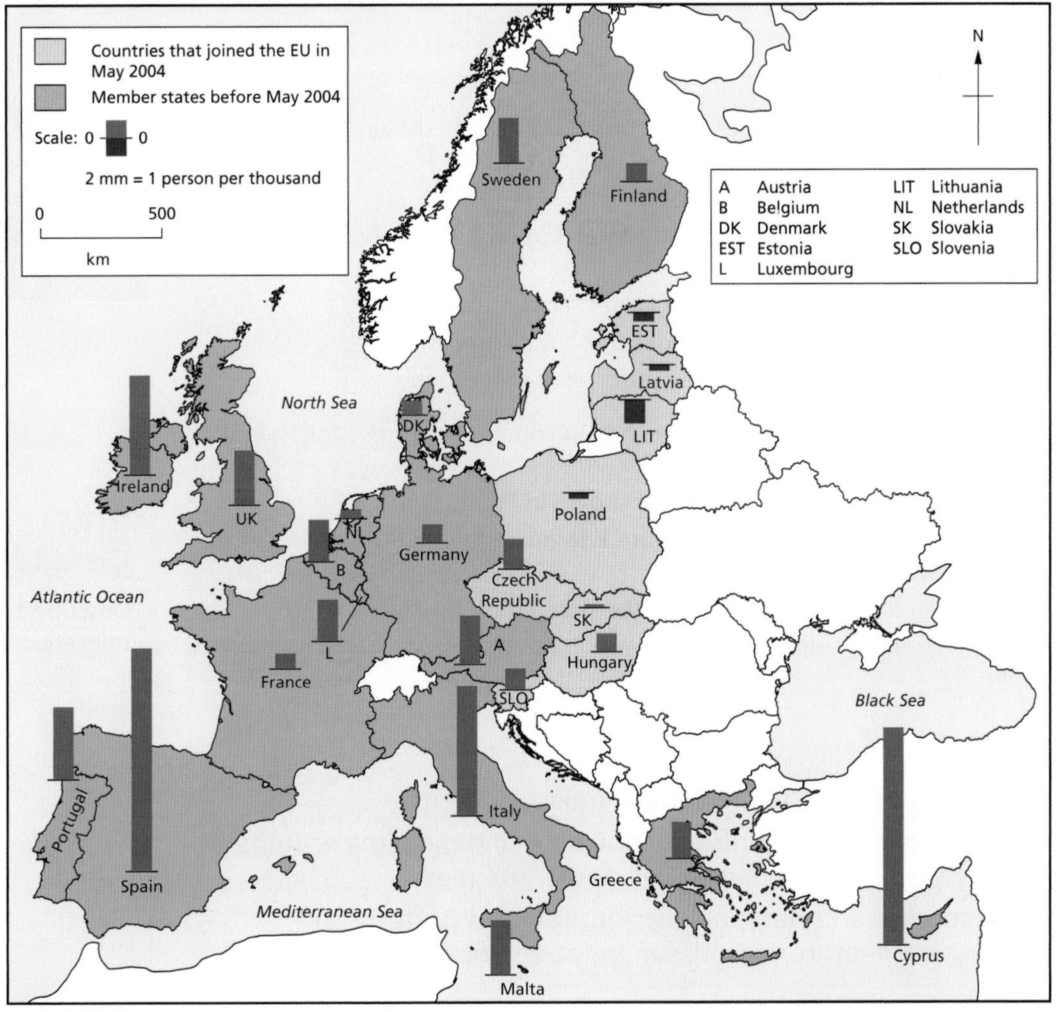

Figure 3 Migration change 2004–2005 in the EU

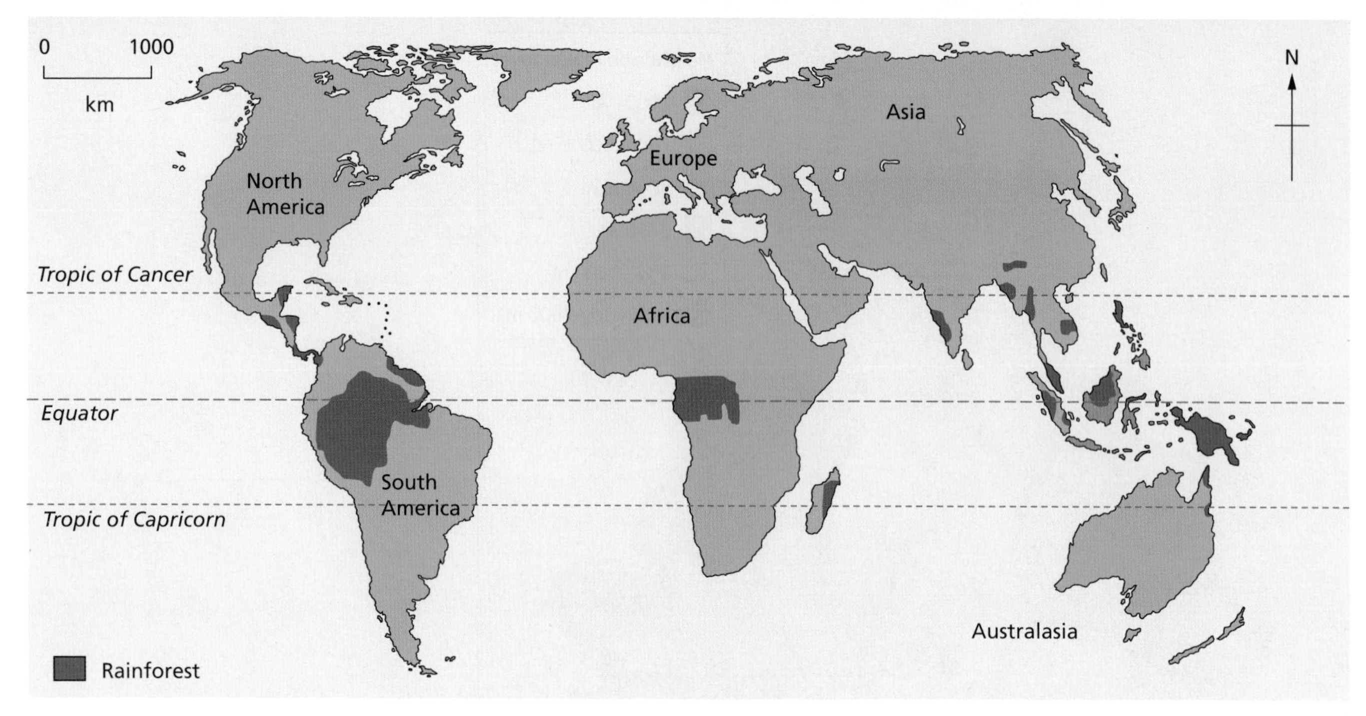

Figure 4 Tropical rainforests of the world

ACTIVITIES

1 a Describe the distribution of tropical rainforests shown in Figure 4.

Answer the following questions:

- Where are most of the rainforests in relation to the named latitude lines?
- Which continent has the largest area of rainforest?
- Which continent does not have any rainforests?
- Name a country where rainforests can be found.

Use questions like those above to help you to answer the following questions:

b Describe the pattern of relief shown in Figure 2.

c What would you add to Figure 2 to make the description easier to complete?

d Compare the distribution of the population in China with the physical features of China shown in Figure 1.

When describing physical and human features always start with the general and finish with the specific.

STRETCH AND CHALLENGE

Describe the pattern of migration shown in Figure 3.

Review

By the end of this section you should be able to:

- recognise and describe distributions and patterns of human features on atlas maps at a variety of scales
- recognise and describe distributions and patterns of physical features on atlas maps at a variety of scales.

Ordnance Survey (OS) maps

Grid references, symbols and direction

Learning objective – to study symbols, grid references and direction.

Learning outcomes
- To be able to recognise symbols.
- To be able to use four- and six-figure grid references accurately.
- To demonstrate an understanding of direction using the eight-point compass.

Scale

Maps are produced at different scales. The scale of the map is how much smaller the map is than the area it represents. On 1:50,000 scale maps, 1 cm on the map is 50,000 cm (500 metres) on the ground. The light blue lines which run horizontally and vertically across the map are called grid lines; they are drawn 2 cm apart. Every map has a scale line which can be used to measure distances; on the skills paper the OS map will also be of the scale 1:50,000.

Symbols

The symbols used on the map represent features on the ground. Many of the symbols give a 'clue' to what they represent. For example, a Post Office is the letter P. You will always be provided with a key but it is a good idea to familiarise yourself with OS maps so that you can interpret them quickly without continual reference to the key.

An OS map symbols quick reference guide is provided below.

Feature	What to look for?
Woodland	Green areas with different types of trees identified
Water	Blue areas
Tourist information	These are symbols in blue
Fields	Usually white areas
Urban areas	Brown/beige shading
Roads	Motorways – blue A roads – green (major routes) and red (main roads) Secondary roads (B) roads – orangey/yellow Small side roads – yellow
Contours (showing height above sea level)	Thin brown lines. They are at 10 m intervals with every 50 m in a darker colour
Spot heights	These are a number next to a dot. They show the height of the land at the point where the dot is
Triangulation pillars	These are a number next to a symbol and also show the height at that point

Grid references

You will need to be able to locate features using four- and six-figure grid references. Four-figure grid references locate a square on the map and are usually used for large features such as an area of settlement or woodland. Six-figure grid references locate a particular point on the map such as a Post Office.

How to find a grid reference

Study the grid in Figure 5a. It has grid lines 2 cm apart and a number of symbols on it. To find the position of the Post Office at 5885:

- Go along the bottom of the grid until you find the number 58.
- These are the first two numbers of the four-figure reference. The line will always be to the left of the symbol. Or if you were on the ground following the map it would be the last grid line you walked over!
- Go up the side of the grid until you reach 85.
- These are the third and four numbers of the grid reference. The line will always be below the symbol you are locating. Or if you were on the ground following the map it would be the last grid line you walked over!
- The symbol will be in the square right and above where these two lines meet.
- If you have been given a six-figure grid reference, the third and sixth numbers refer to the exact position of the symbol, for example 588857.
- You should complete the four-figure part of the reference to find the line and then divide the square into tenths in your mind, in this case, 8/10s as the reference is 588.
- Then complete the second part of the reference in the same way – 857.
- It is important to remember that for features such as a school, you will be given the word 'Sch' near a building. It is the centre of the building that is used for the six-figure grid reference.

Both higher and foundation tier candidates should be able to use six-figure grid references.

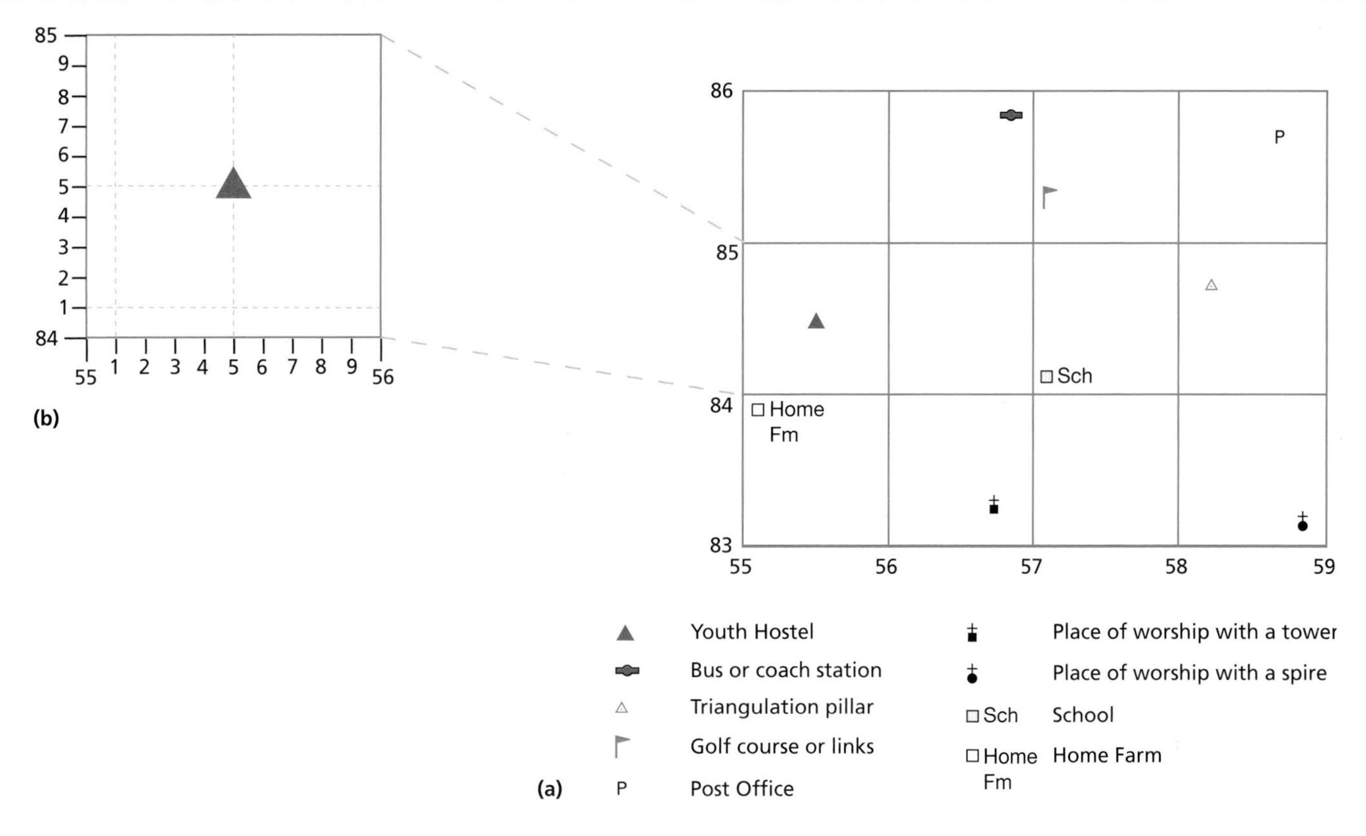

Figure 5 How to find a grid reference

How to give a grid reference

Study the grid in Figure 5a. It has lines 2 cm apart and a number of symbols on it. To give the position of the youth hostel:

- Give the number of the line immediately to the left of the symbol – it is 55.
- Give the number of the line immediately below the symbol – it is 84.
- Therefore the four-figure grid reference for the youth hostel is 5584.
- If a six-figure grid reference is required, you should divide the grid square in your mind into tenths as shown in Figure 5b. (This is double the size of the original square on the grid.) Each of the 1/10s is 2 mm because one grid square is 2 cm on the original grid.
- The youth hostel is halfway across, 5/10s of the way or 555.
- The line below is 84. Again split the grid square into tenths, going upwards this time. The youth hostel is halfway up, 5/10s of the way or 845.
- The six-figure grid reference is written 555845.

Compass directions

Map directions are given using the standard eight-point compass as shown in Figure 6. In an exam, when an OS map is used, north will be taken as being at the top of the map following the grid lines.

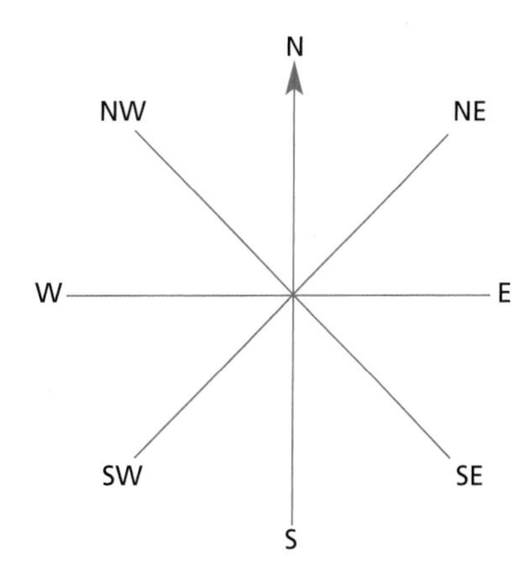

Figure 6 An eight-point compass

Compass directions are often used with oblique photographs.

ACTIVITIES

1 Copy and complete the following table using the key for OS maps on the back cover of the *Tomorrow's Geography* textbook.

Symbol	Feature
TH	
	Tunnel
	Non-coniferous woodland
CH	

2 Copy and complete the table below using Figure 5a and 5b.

Symbol	Key	Four-figure reference	Six-figure grid reference
P			
△			
	Golf course or links		
☐ Home Fm			
	Bus or coach station		
☐ Sch			

Warkworth map extract (page 44)

3 Copy and complete the table below by drawing the symbol for flat rock found in grid square 2511.

Feature	Symbol
Flat rock	

4 Who owns the land in grid square 0825?

5 What type of woodland is found in 2007?

6 Identify three services in the village of Shilbottle in grid square 0819.

7 Complete the table by identifying five tourist features and their grid references found in Alnwick (north-west of the map extract). The first one has been done for you.

Tourist feature	Grid reference
Information	187133

8 Which direction is Alnwick from Alnmouth?

9 Which direction is Warkworth from Alnmouth?

10 Which direction is Shilbottle from Lesbury?

Grid references and directions are often used relating to the location of photographs or features on a map.

STRETCH AND CHALLENGE

1 Look at Figure 6 in Chapter 1, Basic Skills (page 6). The photograph was taken at grid reference 267048. In which direction was the camera pointing?

2 Look at Figure 5 in Chapter 1, Basic Skills (page 6). Which direction was the camera pointing when the aerial photograph was taken?

Review

By the end of this section you should be able to:

- recognise symbols on a map
- use four- and six-figure grid references accurately
- demonstrate an understanding of direction using the eight-point compass.

Straight line and winding distances

Learning objective – to study how to measure distances on a map.

Learning outcomes
- To be able to measure straight line distances on a map.
- To be able to measure curved distances on a map.

Measuring straight line distances

When measuring straight line distances, it is useful to remember that on a 1:50,000 OS map the grid lines are always 2 cm apart, which represents 1 km on the ground. This can make measuring straight line distances very easy.

How to measure a straight line distance between two points such as two railway stations
- Use a ruler or the edge of a piece of paper.
- Mark on the piece of paper the location of the two points or take the measurement between the points with the ruler.
- Measure this distance against the scale line on the map or calculate the distance using the scale of 2 cm = 1 km.

For example, on the OS map of Swanage on page 43, the straight distance from the station at Corfe Castle to the station at Swanage is 14.6 cm or 7.3 km.

Measuring curved (winding) distances

Measuring the distance along a curved or winding route such as a road is more complicated. This can be done using either a piece of string or by splitting the route into straight sections. The easiest way of measuring the distances along a winding route is with a piece of string.

How to measure a route (winding distance such as a railway line, road or river) using a piece of string
- The end of the piece of string should be placed at the beginning of the route.
- The string should be laid along the route following the curves as accurately as possible.
- The end of the route should be marked on the piece of string.
- The length of the string which equals the route can then be measured against the scale on the map or a calculation can be done on the basis that 2 cm = 1 km.

ACTIVITIES

Wensleydale map extract (page 46)

1. Calculate the length of the River Ure from the bridge at Hawes (8790) to the bridge at Bainbridge (9390).
2. What is the approximate length of the A road between 875898 and 934904?
3. Which is the longest route?
4. What is the difference in kilometres?
5. How far is it from one of the sources of Bardale Beck at 865838 to its confluence with Raydale Beck at 911861?

Swanage map extract (page 43)

6. What is the length of the B3069 between its two junctions (9681 and 0078) with the A351?
7. Which is the furthest distance, travelling between these two points on the A351 road or the B3069 road?
8. How much further is it?

How to measure a route (winding distance such as a railway line, road or river) using straight sections

- The route to be measured should be split into a number of straight sections.
- The point chosen for the end of one straight section and the beginning of another is usually where the route bends.
- All of the straight sections can then be measured and the total distance converted into kilometres using the scale.
- The more straight sections that the route is broken down into, the more accurate will be the final measurement.

For example, Figure 7 is a section of the railway line on the OS map of Swanage on page 43. Notice how the line has been broken into straight sections.

To make measuring distances easier always have a piece of string in your pencil case.

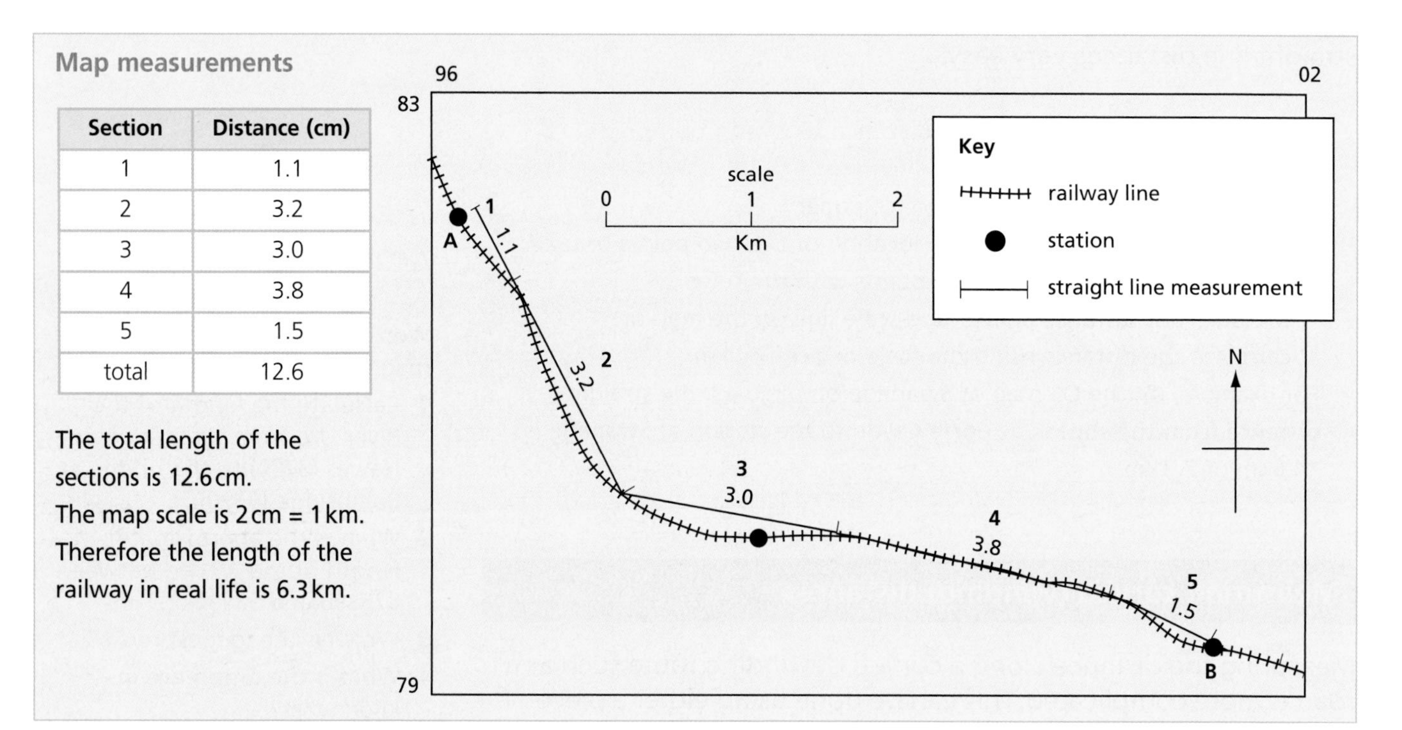

Map measurements

Section	Distance (cm)
1	1.1
2	3.2
3	3.0
4	3.8
5	1.5
total	12.6

The total length of the sections is 12.6 cm.
The map scale is 2 cm = 1 km.
Therefore the length of the railway in real life is 6.3 km.

Figure 7 Measuring distance on an OS map

STRETCH AND CHALLENGE

1. There are three ways to get from Corfe Castle to Swanage; two by road (either the A351 or the B3351) and one by railway. Which is the shortest route?
2. Suggest two reasons for taking the longest route to do the journey.

Review

By the end of this section you should be able to:

- measure straight line distances on a map
- measure curved distances on a map.

Cross-sections

Learning objective – to study how to construct, interpret and annotate cross-sections.

Learning outcomes

- To be able to construct cross-sections.
- To be able to interpret and annotate cross-sections.

A cross-section shows the variations in relief along a chosen line. It is a graph which shows distance along the *x*-axis (horizontal) and height on the *y*-axis (vertical). When drawing a cross-section, the scale used on both axes must be chosen carefully to show a true representation of the area.

How to construct a cross-section along the line X–Y in Figure 8

- Place the edge of a piece of paper along the line X–Y in diagram A.
- Mark the heights of the contour lines where the paper crosses them. This is usually done by putting a small line on the piece of paper and noting down the height at that point. Also mark the spot height which shows the top of the hill.
- Remove the piece of paper with the location of the contour lines and the heights marked on it. Place it on a piece of graph paper. Work out an appropriate scale for the *y*-axis and draw it on the graph paper. Mark points on the graph at the correct height and at the correct location as shown on diagram B. Point A for example is marked at 200 metres.
- You may draw lines up to the points as shown on diagram B and C but it is perfectly correct to just draw the crosses at the correct point.
- Join up the crosses to show the shape of the land. In this case it is a hill with a steep slope and a less steep slope.

Diagram A

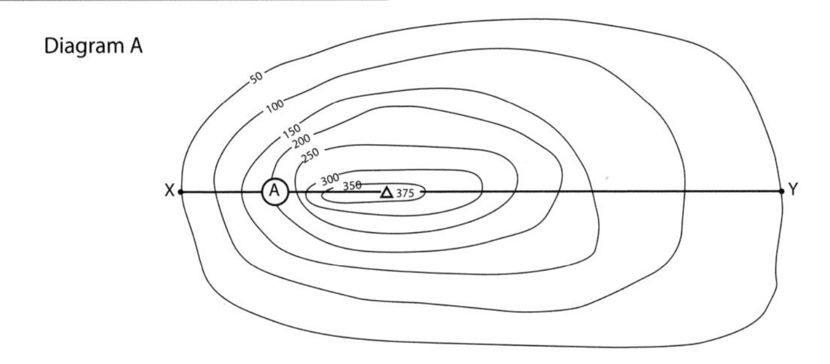

Diagram B

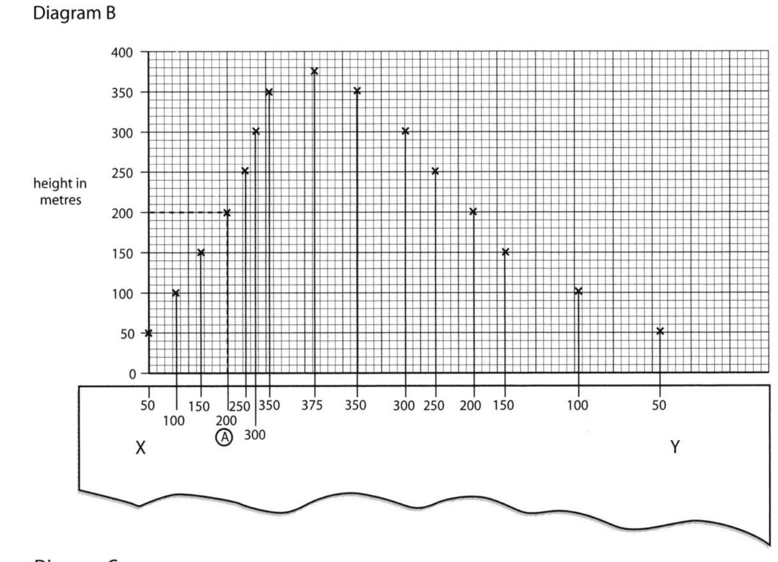

Diagram C

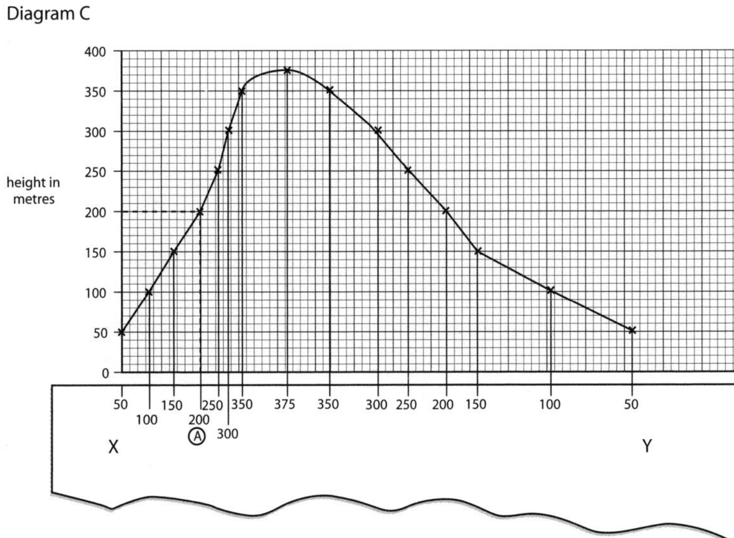

Figure 8 Constructing a cross-section

Interpreting cross-sections

Cross-sections show the shape of the land. If the land is steep, the contours will be close together on the map and the line on the graph paper will increase quickly. If the land slopes gradually, the contours will be further apart and the line on the graph paper will increase more slowly. This can be seen in Figure 8. The first slope to the left of diagram A is steep with the contours on diagram A being quite close together. The contours to the right of diagram A are far apart depicting a gentle slope that is reflected in the line of the graph on diagrams B and C.

Exam Tip

You could be asked to complete a cross-section or annotate it. This means add labels which describe or briefly explain the feature.

Annotating cross-sections

Cross-sections show many physical features. As stated above, the most important feature is the shape of the land. Other physical features such as rivers can also be marked. Human features such as settlements and roads can also be located on a cross-section with explanatory points being made.

ACTIVITIES

Wensleydale map extract (page 46)

1. a Draw a cross-section from the 560 m contour line at Green Side (865851) to the 490 m contour line at New Close (884844).
   b Label the two rivers.
   c Mark on the cross-section where you think the photograph shown in Figure 9 was taken.
   d In which direction do you think the photograph was taken?
   e Justify your answer with reference to the photograph and the map.

Figure 9

2. Draw a cross-section from spot height 511 m in grid square 8987 to spot height 379 m in grid square 9487.
3. Label four secondary roads less than 4 m wide, three streams, Semer Water and two areas of woodland.

STRETCH AND CHALLENGE

Using the Wensleydale map extract, draw a cross-section from spot height 614 m in grid square 8786 to spot height 535 m in grid square 8892. Label the steepest slopes, the River Ure and the A684.

Review

By the end of this section you should be able to:

- construct cross-sections
- interpret and annotate cross-sections.

The distribution and pattern of physical and human features

Learning objective – to study the distribution and patterns made by human and physical features on OS maps.

Learning outcomes

- To be able to recognise and describe the distribution and pattern made by physical features such as slopes (contour patterns), rivers and their valleys and different types of vegetation.
- To be able to recognise and describe the distribution and patterns made by human features such as land use (settlement patterns) and communications.

There are a number of physical and human features which you could be asked to recognise on OS maps and to describe the patterns that they produce. It is important that you understand what the question expects you to write about. The physical features that you could be asked about are:

- **Relief** – the shape of the land. This can be seen by looking at the distance between the contours and the pattern that the contours show on the map. The height difference between each contour line (the vertical interval) is 10 m on a 1:50,000 map. Every 50 m the contour line is a slightly darker brown and has the height marked on it as shown in Figure 10 on page 24. When writing about the relief you should use actual figures from the map, not terms such as highland, lowland or steep slopes. These are meaningless without the actual height or the rate at which the height is increasing or decreasing.
- **Rivers and their valleys** – the skills you require are to be able to:
  - recognise the presence or absence of rivers or lakes in an area
  - recognise the patterns that rivers make in an area
  - recognise the direction that a river is flowing
  - see the influence of human activity on rivers and their valleys
  - describe a river and its valley
  - compare river valleys on a map extract.
- **Vegetation** – you will need to be able to recognise the type of vegetation, for example, woodlands, orchards and parks and describe their distribution on a map. These can be seen in the land features section of the key.

Contour patterns	Commentary
50	This is a uniform slope. The contours decrease evenly. The distance between the contour lines is the same all the way down the slope.
100 50	This is a convex slope. The contour lines are closer together at the bottom of the slope. This means that the height is increasing quickly because the slope is steep. The lines gradually become further apart towards the top of the slope. This means that the height is increasing more slowly and that the slope is more gentle.
50	This is a concave slope. The contour lines are further apart at the bottom of the slope. This means that the height is increasing slowly and that the slope is gentle. The lines gradually become closer together towards the top of the slope. This means that the height is increasing more quickly and that the slope is steeper.
100	This is a V-shaped valley. The distance between the contours is regular, meaning that the valley sides will have an even slope. There is no flat land by the side of the river. This is shown by the presence of contour lines right next to the river. The V of the contour always points upstream. In this instance the river is flowing from the top to the bottom of the page.
50 50	This is a U-shaped valley. The distance between the contours is regular, meaning that valley sides will have an even slope. The contours increase quickly, meaning that the sides of the valley are steep. There is an area of flat land by the side of the river shown by the lack of contour lines. The contour lines do not cross the river for many kilometres; it is therefore very difficult to ascertain the direction of flow. On an OS map other clues would be the presence of tributaries and the river becoming wider as it moves downstream.

Figure 10 Basic contour patterns of slopes and valleys

Exam Tips

- **Remember that contour lines never touch or cross.**
- **The V of the contours across a river always points upstream.**
- **Contour lines which are close together mean steep slopes.**
- **Contour lines which are far apart mean gentle slopes.**

The human features that you could be asked about are:

- **Land use** – this would include urban areas and settlement patterns in rural areas. Typical questions would ask about the settlement distribution of an area and the human or physical factors which might have influenced the pattern.
- **Communications** – this would include routes such as roads and railway lines. You might be asked to comment on the pattern shown by the communication links or explain how relief has influenced the pattern.

How to describe the pattern or distribution of settlement on a map

- Begin with a general statement about where settlement is located on the map such as 'in the south-east'.
- Then go into greater detail, such as 'the settlement is along the lines of communication and at the bottom of a steep slope'. This mentions both a physical and human feature although this is not essential.
- You should then be more specific and include in the answer the name of the road and the height of the land.

For example, there follows a description of the settlement pattern on the Wensleydale OS map (page 46). The majority of the settlement can be found in a band across the centre of the map extract. The settlements are close to the River Ure. The largest settlement is Hawes in grid square 8789. The settlements are mainly found on the south side of the valley, an exception being Askrigg in grid square 9491.

STRETCH AND CHALLENGE

Explain the communication links shown on the Cambridge OS map extract on page 42.

ACTIVITIES

Looe map extract (page 45)

1 Describe the distribution of woodland.

2 Describe the distribution of communication links in the area.

Swanage map extract (page 43)

3 Describe the distribution of settlement.

4 Woodland is being managed in the area of the map north of grid line 82. How does the map show that the woodland is being managed?

5 How have communication links been affected by the relief of the area?

Review

By the end of this section you should be able to:

- recognise and describe the distribution and pattern made by physical features such as slopes (contour patterns), rivers and their valleys and different types of vegetation
- recognise and describe the distribution and patterns made by human features such as land use (settlement patterns) and communications.

Site, situation and shape of settlements

Learning objective – to study the site, situation and shape of settlements.

Learning outcomes

- To be able to describe and identify the site and situation of settlements.
- To be able to describe and identify the shape of settlements.

Gap town

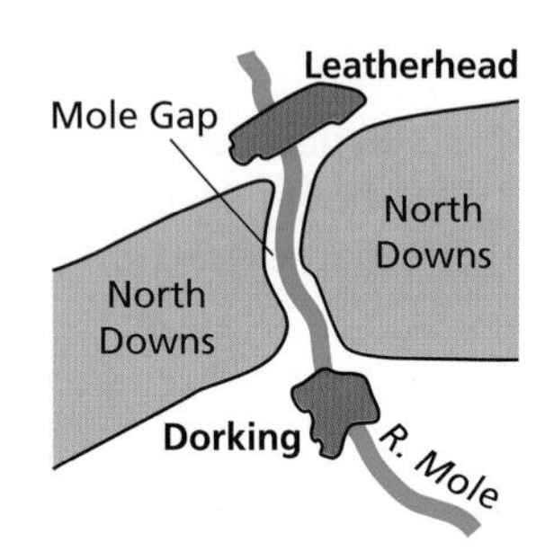

This shows the gap towns of Leatherhead and Dorking. They have been sited either side of a gap in the North Downs. These are a range of hills that stretch across the south-east of England forming a barrier to communication routes. Originally there may have been a castle here to control the gap. (For example, Corfe Castle on Swanage map.)

Defensive site

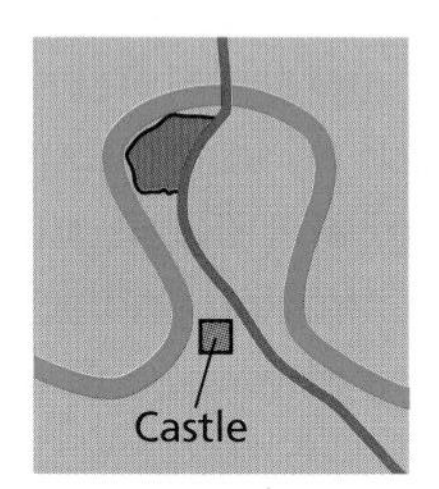

This is a defensive site. The settlement is sited inside a meander bend to give protection to the settlement. It is surrounded on three sides by water. Often, as in this case, the 'open' end was defended by a castle. Other sites for defensive settlements were on the top of hills. This allowed them to be prepared for invaders.

Lowest bridging point

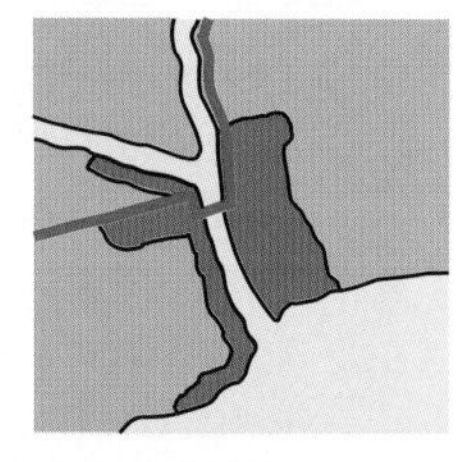

This site is at the lowest bridging point of a river. This means the last place that a river could either be 'forded' (walked across) or 'bridged' before the sea.

Coastal town

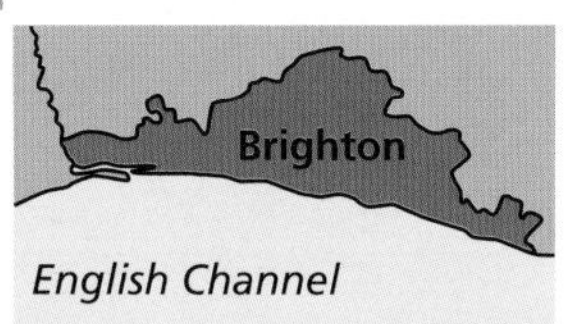

This is a coastal town. There are many towns and cities around the UK which are sited on the coast. Many of them were sited originally for either communication links due to the harbour facilities or, as in the case of Brighton, for tourism.

Spring-line settlements

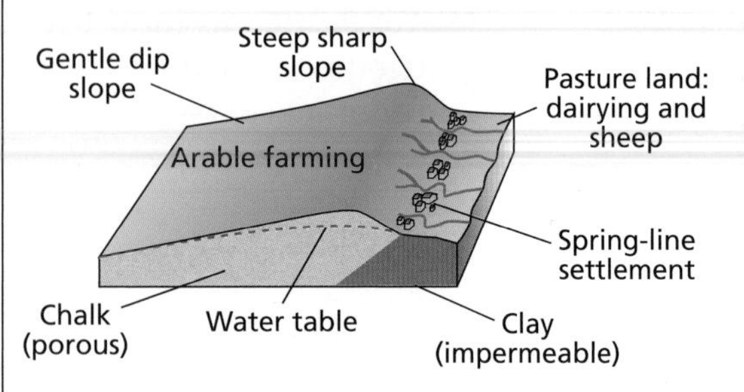

This site shows spring-line settlements. The site is on a spring at the base of a line of hills. The spring provides a water supply and the hills offer protection from the wind.

Figure 11 Settlement sites

What is the site of a settlement?

This is the land that the settlement is actually built on. If you are describing the site of a settlement on an OS map, you should describe a number of human and physical features. Try to remember **SHAWL**:

- **S**helter from strong winds and storms
- **H**eight above sea level
- **A**spect, the way that the slope faces
- **W**ater supply
- **L**and that the settlement is built on, such as above the floodplain, fertile land, type of slope.

What is the situation of a settlement?

This is the settlement's position in relation to its surroundings. When you are describing the situation of a settlement on an OS map, you should describe the human and physical features around it. Try to remember **PARC**:

- **P**laces
- **A**ccessibility
- **R**elief
- **C**ommunications.

What is the shape of a settlement?

This is the arrangement of houses within the settlement; the form that the settlement takes. On an OS map you should be able to identify the following shapes (see Figure 12):

- nucleated, where the buildings are grouped together. They often form at crossroads or around a village green
- linear settlements have their buildings either side of a main road, along a valley or the coast
- dispersed settlements have individual buildings spread out around an area, there is no obvious centre.

ACTIVITIES

Looe map extract (page 45)

1. Describe the site of Herodsfoot (2160).
2. Compare the shape of Herodsfoot with the shape of Duloe (2358).

Swanage map extract (page 43)

3. Describe the situation of Swanage.

You might be asked to draw a sketch of a settlement and annotate it to explain its site.

STRETCH AND CHALLENGE

Explain the situation of Looe.

Review

By the end of this section you should be able to:

- describe and identify the site and situation of settlements
- describe and identify the shape of settlements.

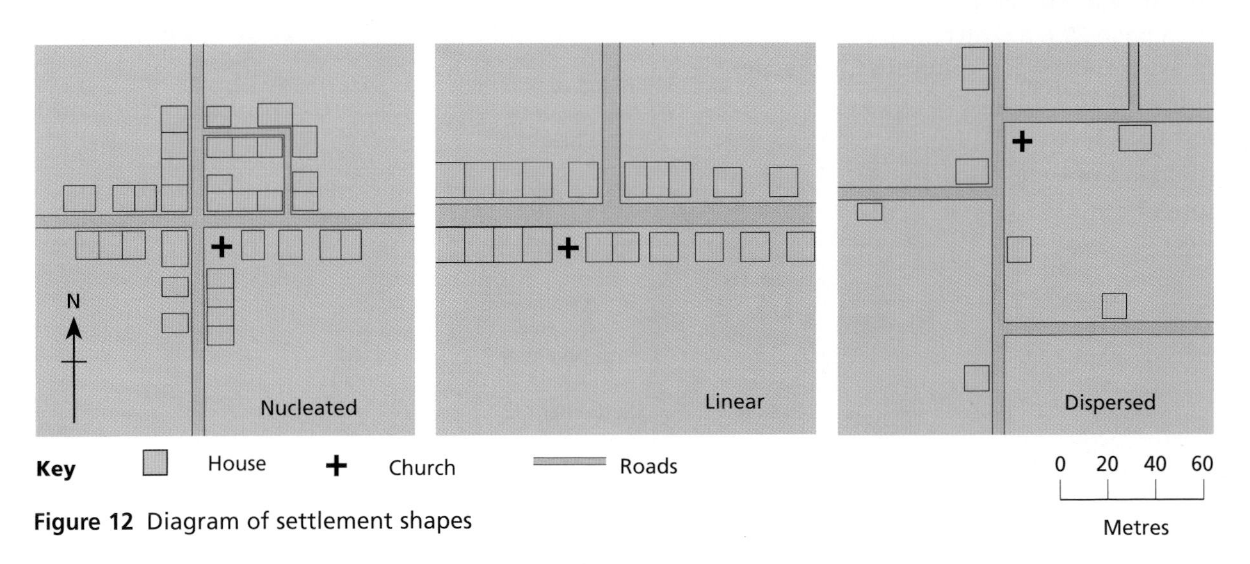

Figure 12 Diagram of settlement shapes

Human activity from map evidence

Learning objective – to study map evidence of human activity including tourism.

Learning outcomes

- To be able to interpret map evidence of human activity.
- To be able to interpret map evidence of tourism.

Maps show a large amount of human activity, the extent of which is shown by looking at the key. Communication routes such as roads and railways with their associated features such as tunnels and level crossings (LC) are shown. Other public rights of way such as footpaths and cycle networks are also shown on maps.

The section on the OS map key called 'Land features' contains a lot of information about the location of other human activity, for example windmills and wind generators. It also indicates who owns some of the land on the map, such as the National Trust.

Maps show settlements but not the number of people who live in them, although their actual size is shown by the area they cover. Some human activity is shown by the services available in settlements, such as churches, Post Offices and public houses.

Tourist information has its own section in the key on OS maps, and many symbols are used to depict the variety of tourist information that is available on maps. As well as the information in the key, it is also important to remember that castles (in the antiquities part of the key) are also tourist attractions.

ACTIVITIES

Looe map extract (page 45)

1 Duloe has five services. State what the services are by drawing the appropriate symbol.

2 Figure 13 on page 29 is a sketch map of part of the Looe map extract. Copy and complete the sketch map with the following features:
   a the West Looe River
   b the settlement of Looe
   c the route of the A387
   d 100 m contour line.

3 Copy and complete the table using the sketch map and the OS map extract of Looe to help you.

Place	Feature on map
Island A	
Spot height B	
Tourist attraction C	
Tourist attraction D	
Road number E	
Road number F	
Settlement G	
Type of woodland H	
Farm I	

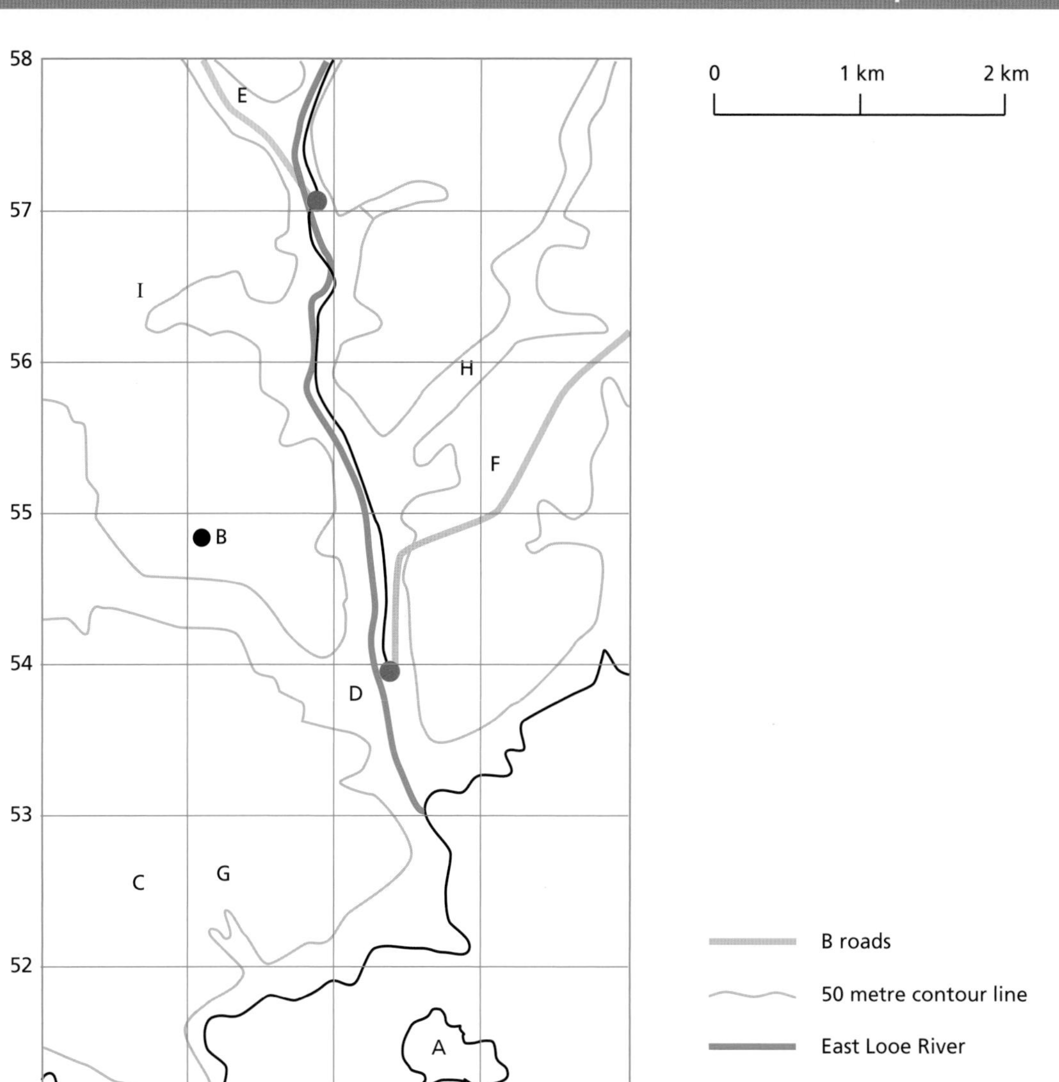

Figure 13 An incomplete sketch map of part of the Looe extract

Human activity is likely to be examined by using a sketch map. On exam papers it is better to shade rather than use different colours on sketch maps.

STRETCH AND CHALLENGE

Explain the presence of so many tourist information features on the Looe map extract on page 45.

Review

By the end of this section you should be able to:

- interpret map evidence of human activity
- interpret map evidence of tourism.

Use maps in association with photographs, sketches and directions

Learning objective – to study the use of maps in association with photographs, sketches and directions.

Learning outcomes

- To be able to interpret maps and relate them to photographs.
- To be able to interpret maps and relate them to sketches.
- To be able to follow directions on a map.

Many exam papers have questions which require you to be able to use the OS map with a photograph. The types of questions you could be asked are:

- recognition of certain features which have been identified by a letter on the photograph
- the direction the photograph was taken
- a comparison of features that can be seen on the map and not on the photograph and on the photograph and not on the map
- the location of where the photograph was taken.

In order to be able to use a photograph with a map, you will have to orientate the photograph so that the features on the photograph are in the same position as those on the map. The map always has north at the top, a photograph may not! The way to do this is to look for an important feature on the photograph, such as the shape of the coastline on the Swanage map (page 43) and Figure 14. Then look at the town shown by the brown shading on the map and the houses on the photograph. It is obvious that the photograph is taken from the north of the map facing south. The exact location is worked out by looking at the fine detail of the photograph in relation to the map; in this case it would be the angle of the coastline, the position of the pier and the headland in the distance. Figure 14 has been annotated with the answers to questions that you might be asked when using a map with a photograph.

Other examination questions may require you to use the OS map to complete a sketch. There also could be sketches drawn with features labelled that you then have to identify using the map. There is an example of this on page 29 where there is a sketch of the Looe map. This has been covered in more detail in the Basic Skills chapter.

There may also be questions which ask you to give directions from one place to another or to follow directions identifying features that you pass on the way. You could also be asked to identify the best route between two places.

Figure 14 A photograph of Swanage Bay taken from 040814 facing south

If you are using a photograph it is important to orientate it before you start to answer the questions, so that the map and the photograph are facing the same direction with north at the top.

ACTIVITIES

Looe map extract (page 45)

1 Study Figure 15. It is an aerial photograph of Looe.

a In which direction was the camera pointing?

b Copy and complete the table below using the photograph and the OS map of Looe on page 45 to help you.

Letter on photograph	Grid reference	Feature
V	Six figure =	Number of road =
W	Four figure =	Name of woodland =
X	Six figure =	Identify symbol =
Y	Four figure =	Name of river =
Z	Six figure =	Identify symbol =

c State two features that can be seen on the photograph but not on the map.

d State two features that can be seen on the map but not on the photograph.

Swanage map extract (page 43)

2 A family are on holiday in the Swanage area, shown on the map on page 43. They are staying at Corfe Castle and want to go out for the day visiting Studland and Swanage. Their route is described below. Copy and complete the route. Use the OS map extract of the Swanage area and some of the terms in the box to help you.

Cemy	038826	A351	Old Harry	Ulwell
Rollingham Farm		826038	B3069	A315
B3351	Ballard Point	Brenscombe Farm		

They leave Corfe Castle on the ...... towards Studland. After they have travelled for 2 km, they pass ...... on their right. They arrive in Studland and park the car in the car park closest to the Post Office at grid reference ...... The family walk in an easterly direction to ...... in grid square 0582. They leave Studland and travel to Swanage. In the village of ...... they pass a camp site and a caravan site. At the end of the day they return to their hotel. They leave Swanage on the ...... to Corfe Castle.

Looe map extract (page 45)

3 A family wish to visit Looe and Polperro. They are staying in Herodsfoot. Describe the route they would take from their hotel to visit both places.

Figure 15 Aerial photograph of Looe

STRETCH AND CHALLENGE

Another family are on holiday in Looe and want to go to Plymouth for the day. Plan their route.

Review

By the end of this section you should be able to:

- interpret maps and relate them to photographs
- interpret maps and relate them to sketches
- follow directions on a map.

Sample Examination Questions

Cambridge map

Higher tier

1 Study Figure 16. It is an aerial photograph of an area of housing in Cambridge. Also study the OS map on page 42.

a A number of grid squares are shown in the photograph. Choose one of them from the following: 4460, 4657, 4856, 4458. **(1 mark)**

b What is feature X? **(1 mark)**

c Give the six-figure grid reference for feature Y. **(1 mark)**

d What is the number of the road marked Z? **(1 mark)**

e Describe the pattern of housing identified by the pink square on the photograph. **(3 marks)**

2 Figure 17 is a photograph which was taken above Cherry Hinton (4856):

a Compare the housing patterns shown in Figures 16 and 17. **(3 marks)**

b Suggest one reason for these differences. **(1 mark)**

3 Copy and complete the table by identifying features which can be seen on the map and not on the photograph and features that can be seen on the photograph and not on the map. **(2 marks)**

	On map not on photograph	On photograph not on map
Feature 1		
Feature 2		

4 Describe the distribution of 'park and ride' sites shown on the map. **(3 marks)**

Foundation tier

1 Look at Figure 16. It is an aerial photograph of an area of housing in Cambridge. Also look at the OS map on page 42.

a A number of grid squares are shown in the photograph. Choose one of them from the following: 4460, 4657, 4856, 4458. **(1 mark)**

b What is the railway feature identified by the X? **(1 mark)**

c Give the six-figure grid reference for the college identified by feature Y. **(1 mark)**

d What is the number of the A road marked Z? **(1 mark)**

e Describe the pattern of housing shown in the pink square on the photograph. Copy out the three following statements which best describe the housing identified by the pink square:

- The houses are in cul-de-sacs.
- The housing is in a grid iron pattern.
- The streets have curves.
- The houses are in blocks.
- There are lots of deadends.
- The streets are in a line. **(3 marks)**

2 Figure 17 is a photograph which was taken above Cherry Hinton (4856):

a State two differences between the housing types shown in Figures 16 and 17. **(2 marks)**

b Suggest one reason for these differences. **(1 mark)**

3 Copy and complete the table by identifying features which can be seen on the map and not on the photograph and features that can be seen on the photograph and not on the map. **(2 marks)**

	On map not on photograph	On photograph not on map
Feature 1		
Feature 2		

4 Complete the sentences to describe the distribution of the 'park and ride' sites shown on the map. Use some of the terms in the box below:

railway	outskirts	major	top	road
4259	motorway	5942		

The 'park and ride' sites are all on the ...... of Cambridge. They are close to ...... roads that lead into the city. There is a 'park and ride' site in grid square ...... This is close to a ...... junction. **(4 marks)**

Sample Examination Questions

Figure 16 Aerial photograph of housing in Cambridge

Figure 17 Aerial photograph of Cherry Hinton

Sample Examination Questions

Swanage map

Higher tier

1 Study the OS map of Swanage on page 43. A family are on holiday in the area shown on the OS map. They are staying near Corfe Castle in grid square 9582. They wish to go out for the day to visit Studland and Swanage. Answer the following questions about their route and the features they see on the way:

a What is the number of the road which leads from Corfe Castle to Studland? **(1 mark)**

b They park close to the Post Office in Studland. What is the six-figure grid reference of the Post Office? **(1 mark)**

c They walk in an easterly direction to Old Harry. Who owns Old Harry? **(1 mark)**

d The family leave Studland and drive to Swanage. In Ulwell they pass a tourist feature. Which tourist feature do they pass in Ulwell? **(1 mark)**

e They decide to return to Corfe Castle by train. How many stations are there on the Swanage railway? **(1 mark)**

2 Compare the shape of Harmans Cross in grid square 9880 with that of Worth Matravers in grid square 9777. **(3 marks)**

3 Relief can greatly affect the routes taken by roads. Describe one example from the map where a road has been affected by relief. **(2 marks)**

4 Study Figure 18. It is a photograph of Swanage Bay. Draw a sketch of the photograph. Label the following on the sketch (you will also need to refer to the OS map):

a the town of Swanage

b the beach

c the groynes on the beach

d Swanage Bay

e Peveril Point

f the pier. **(5 marks)**

Foundation tier

1 Study the OS map of Swanage on page 43. A family are on holiday in the area shown on the OS map. They are staying near Corfe Castle in grid square 9582. They wish to go out for the day to visit Studland and Swanage. Answer the following questions about their route and the features they see on the way:

a What is the number of the road which leads from Corfe Castle to Studland?

i A351

ii B3351

iii B3069

iv A3351. **(1 mark)**

b They park close to the Post Office in Studland. Which one of the following is the six-figure grid reference of the Post Office?

i 025822

ii 038826

iii 035822

iv 822035. **(1 mark)**

c They walk in an easterly direction to Old Harry. Which one of the following owns Old Harry?

i Harry Thompson

ii National Trust

iii Swanage council

iv National Term. **(1 mark)**

d The family leave Studland and drive to Swanage. In Ulwell they pass a tourist feature. Which tourist feature do they pass in Ulwell? **(1 mark)**

e They decide to return to Corfe Castle by train. Choose how many stations there are on the Swanage railway from the following:

i 3

ii 4

iii 5

iv 6. **(1 mark)**

Sample Examination Questions

2 Complete the sentences to compare the shape of Harman's Cross in grid square 9880 with that of Worth Matravers in grid square 9777. Use some of the terms in the box.

B3351	nucleated	junction	A351	linear

Harman's Cross is a .... village. Most of the houses are either side of the road, the .... The village of Worth Matravers has a more .... shape with the houses being grouped around a ...... **(4 marks)**

3 Study Figure 18. It is a photograph of Swanage Bay. Draw a sketch of the photograph. Label the following on the sketch (you will also need to refer to the OS map):

- **a** the town of Swanage
- **b** the beach
- **c** the groynes on the beach
- **d** Swanage Bay
- **e** Peveril Point. **(6 marks)**

Figure 18 A photograph of Swanage Bay

Sample Examination Questions

Warkworth map

Higher tier

1 Study Figure 19. It is a photograph taken at grid reference 242063.

- a In which direction was the camera pointing? **(1 mark)**
- b Name the river flowing under the bridge. **(1 mark)**

2 Study Figure 20. It is a sketch map of part of the OS map extract on page 44. Trace a copy of the sketch map. Label the following features on the sketch map. You should use the correct symbol and provide a key.

- a Alnmouth station **(1 mark)**
- b Public house **(1 mark)**
- c Post Office **(1 mark)**
- d Complete the road network on the map by adding the A1068, the B1339 and the B1338. **(2 marks)**
- e Describe the course of the River Aln shown on the sketch map. **(4 marks)**

3 A family are on holiday staying at Hermitage Farm in grid square 2406. They would like to visit Warkworth for lunch and then go onto Alnwick castle in the afternoon. Describe their route. **(4 marks)**

Foundation tier

1 Look at Figure 19. It is a photograph taken at grid reference 242063.

- a In which direction was the camera pointing?
  - i north
  - ii south
  - iii east
  - iv west. **(1 mark)**

Figure 19 A photograph of Warkworth

Sample Examination Questions

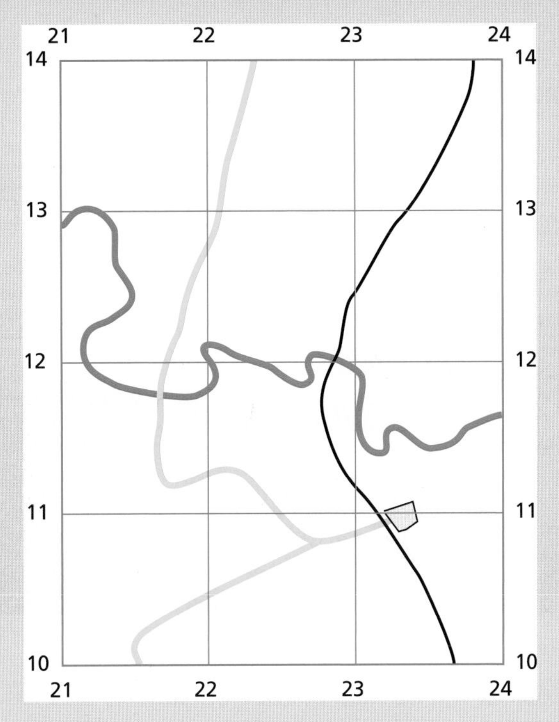

Figure 20 A sketch map of Warkworth

b Name the river flowing under the bridge.

i Warkworth

ii Hermitage

iii Aln

iv Coquet. **(1 mark)**

2 Look at Figure 20. It is a sketch map of part of the OS map extract on page 44. Trace a copy of the sketch map.

a Label the following features on the sketch map. You should use the correct symbol and provide a key:

Feature	Grid square
Alnmouth station	2311
Public house	2311

(2 marks)

b Complete the road network on the map by adding the A1068 and the B1339. **(2 marks)**

c Complete the sentences below to describe the features of the valley of the River Aln shown on the sketch map. Use some of the terms in the box:

A1068	B1339	2112	2212	south	north

A railway line crosses the river in grid square ...... In grid square 2311 the river is crossed by the .... The river flows to the .... of Lesbury. There is a small coniferous woodland to the .... of the river in grid square 2211. **(4 marks)**

3 A family are on holiday staying at Hermitage Farm in grid square 2406. They would like to visit Alnwick Castle. Put the following sentences into the correct order to describe their route. The first has been done for you.

Leave Hermitage Farm, turn left on to the secondary road.

Go straight across the roundabout in grid square 2311, continuing on the A1068.

After approximately 5 km you will then reach the town of Alnwick.

After approximately 1 km arrive at a junction with the A1068.

Continue through the town on the B6341, parking in the car park close to the castle.

Turn left onto the A1068 towards Alnmouth, drive for approximately 5 km. **(5 marks)**

Sample Examination Questions

Looe map

Higher tier

1 Study the OS map of Looe on page 45 and Figure 21. It is a photograph taken in grid square 2552 facing south.

a What is the name of island S? **(1 mark)**

b What is feature T? **(1 mark)**

2 Suggest reasons for the site of Pelynt. **(4 marks)**

3 Study the OS map of the Looe area:

a Draw a cross-section from the triangulation pillar 168 in grid square 2261 to Lametton Mill at 260610. You may use the one in Figure 22 on page 39 to help you.

b Label on the cross-section: East Looe River, B3245, railway line, non-coniferous woodland. **(5 marks)**

4 Study the OS map of Looe and Figure 23. It is a photograph taken in grid square 2552 facing north. There is information on the map that is not on the photograph:

a Label on the photograph three pieces of information that are present on the map but not the photograph. **(3 marks)**

b State one piece of information that is on the photograph and not the map. **(1 mark)**

Foundation tier

Figure 21 A photograph of Looe

1 Look at the OS map of Looe on page 45 and Figure 21. It is a photograph taken in grid square 2552 facing south.

a What is the name of island S?

i Mid Main

ii Hore Stone

iii Looe Island

iv The Ranneys. **(1 mark)**

b What is feature T?

i cliff

ii flat rock

iii spoil heap

iv low water mark. **(1 mark)**

2 Complete the sentences to suggest reasons for the site of Pelynt. Use some of the words in the box below.

spring	wet	dry	lake	above
on	secondary	drinking	dirty	

Pelynt is a …….. point site. It has a …….. in the village which would have supplied …….. water to the villagers. It is …….. the floodplain of the river. It can found where a number of …….. roads meet. **(5 marks)**

Sample Examination Questions

3 Look at the OS map of the Looe area.

a Draw a cross-section from the triangulation pillar 168 in grid square 2261 to Lametton Mill at 260610. You may use the one in Figure 22 to help you.

b Label on the cross-section: East Looe River, B3245, railway Line, non-coniferous woodland. **(5 marks)**

4 Look at the OS map of Looe and Figure 23. It is a photograph taken in grid square 2552 facing north. There is information on the map that is not on the photograph.

a Label on the photograph three pieces of information that are present on the map but not the photograph. Choose three pieces of information from the list below: **(3 marks)**

place of worship with a tower	beacon museum
railway station	Discovery centre
car park	IRB station

b State one piece of information that is on the photograph and not the map. **(1 mark)**

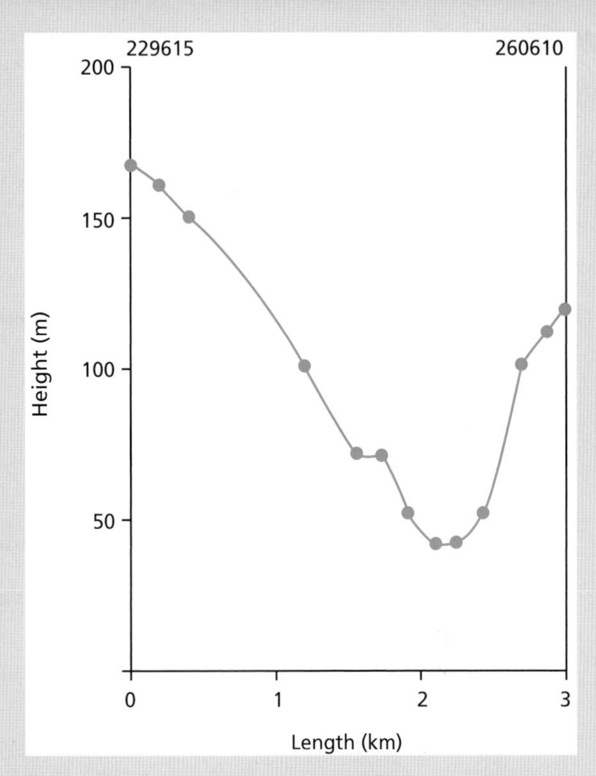

Figure 22 Cross-section from 229615 to 260610

Figure 23 A photograph of Looe

Sample Examination Questions

Wensleydale map

Higher tier

1 Study the OS map of Wensleydale on page 46 and Figure 24. It is a photograph taken at grid reference 876906.

a Name the river shown in the photograph. **(1 mark)**

b In which direction is the river flowing?

i east to west

ii north to south

iii west to east

iv south to north. **(1 mark)**

2 Copy and complete the following table by ticking (✓) in the box if Hawes (8789) and Bainbridge (9390) have that service. **(4 marks)**

Service	Hawes	Bainbridge
Public house		
Place of worship with a spire		
Visitor centre		
Post Office		

3 Most of the settlements in the area are located in the river valleys. Suggest reasons why. Use evidence from the map in your answer. **(4 marks)**

4 Contrast the valley of the River Ure between grid reference 880905 and 900903 with the valley of Cragdale Water between grid reference 919840 and grid reference 911855. **(5 marks)**

Foundation tier

1 Study the OS map of Wensleydale on page 46 and Figure 24. It is a photograph taken at grid reference 876906.

a Name the river shown in the photograph:

i Ure

ii Hardraw Force

iii Bain

iv Brown. **(1 mark)**

b In which direction is the river flowing?

i east to west

ii north to south

iii west to east

iv south to north. **(1 mark)**

2 Copy and complete the following table by ticking (✓) in the box if Hawes (8789) and Bainbridge (9390) have that service. **(3 marks)**

Service	Hawes	Bainbridge
Public house		
Place of worship with a spire		
Post Office		

3 Identify three pieces of evidence which show that tourism is important in this area. **(3 marks)**

4 Most of the settlements in the area are located in the river valleys. Suggest one reason why. Use evidence from the map in your answer. **(2 marks)**

5 Contrast the valley of the River Ure between grid reference 880905 and 900903 with the valley of Cragdale Water between grid reference 919840 and grid reference 911855. Use some of the terms in the box.

west to east	south-east to north-west
south north	steep gentle small large

Cragdale water has a number of …….. bends; the Ure has one big one. The River Ure has a wide, flat valley floor about 0.5 km. The river flows from …….. There is a caravan site to the …….. of the river. Cragdale water has …….. sides and very little flat land in the valley bottom. The river flows from …….. **(5 marks)**

Sample Examination Questions

Figure 24 A photograph of Wensleydale

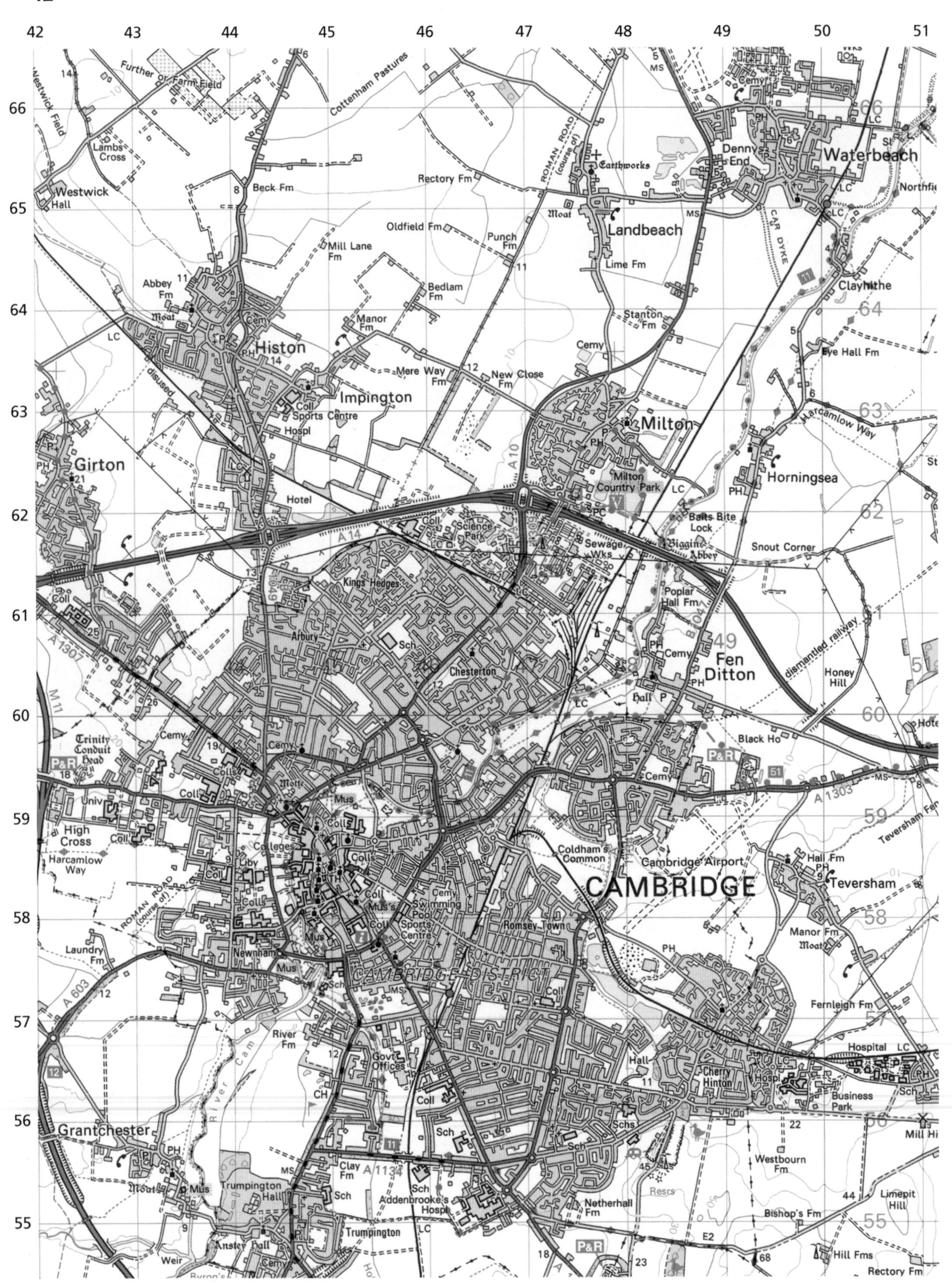

OS map of Cambridge

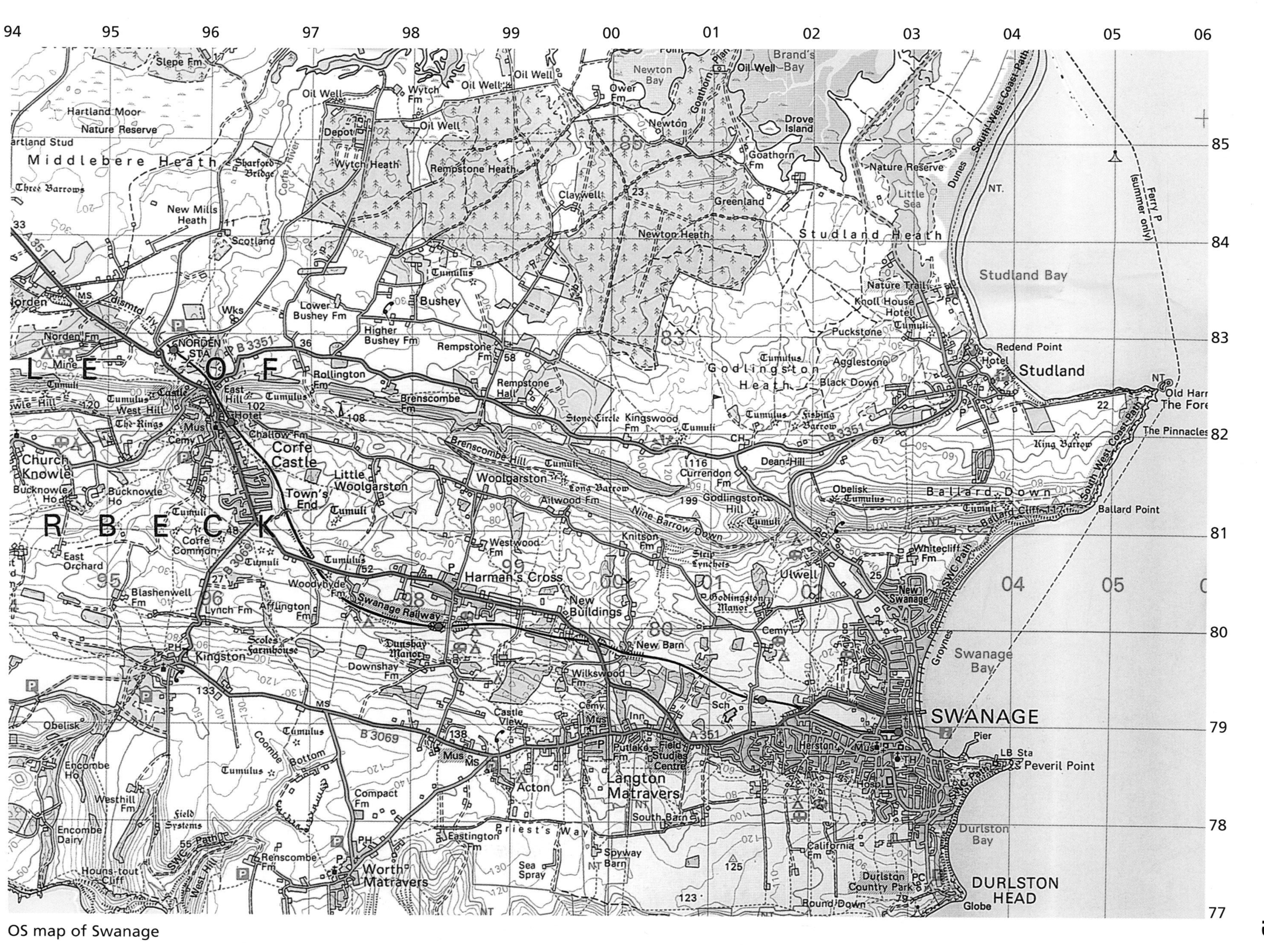

OS map of Swanage

OS map of Warkworth

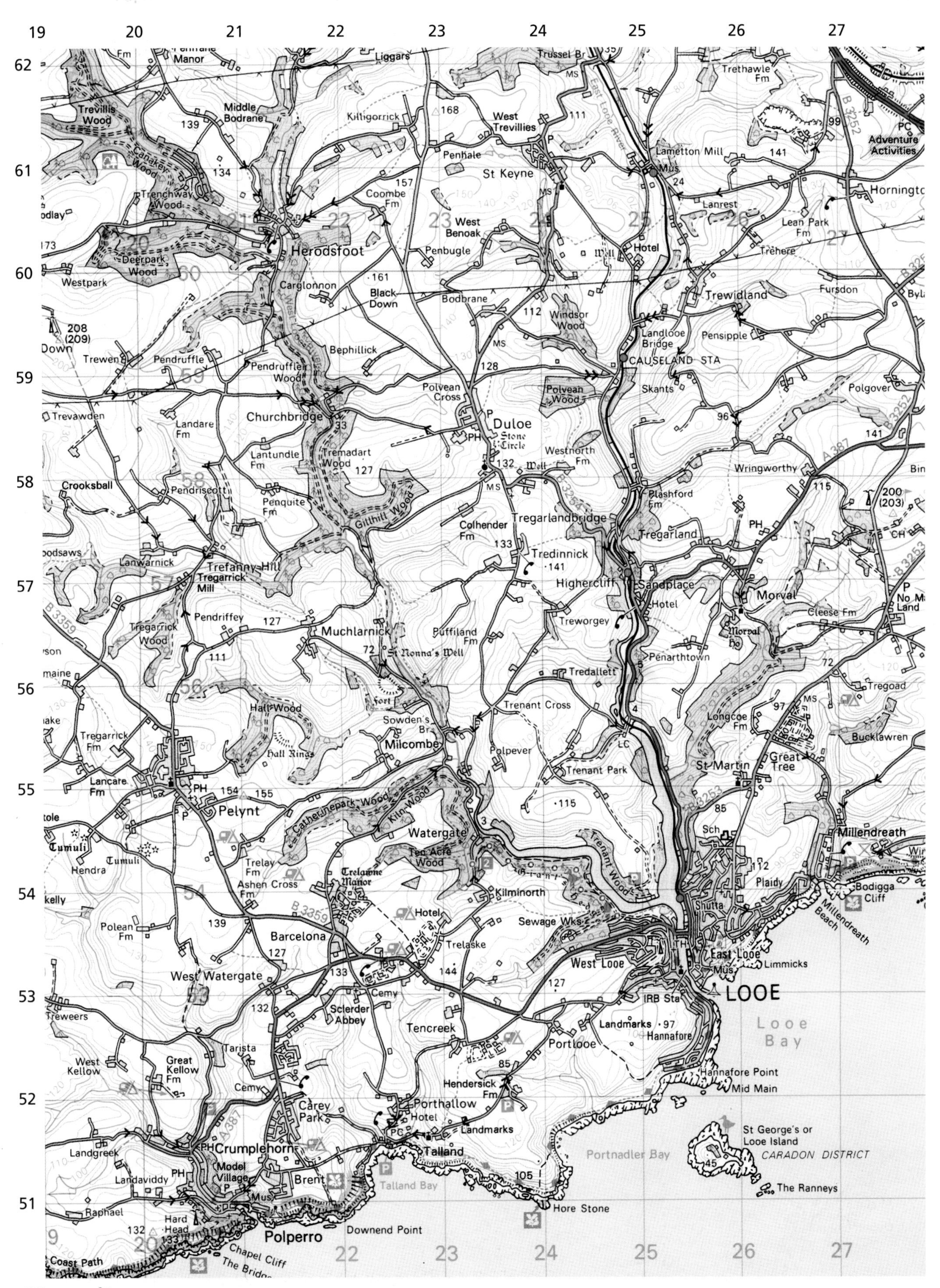

OS map of Looe

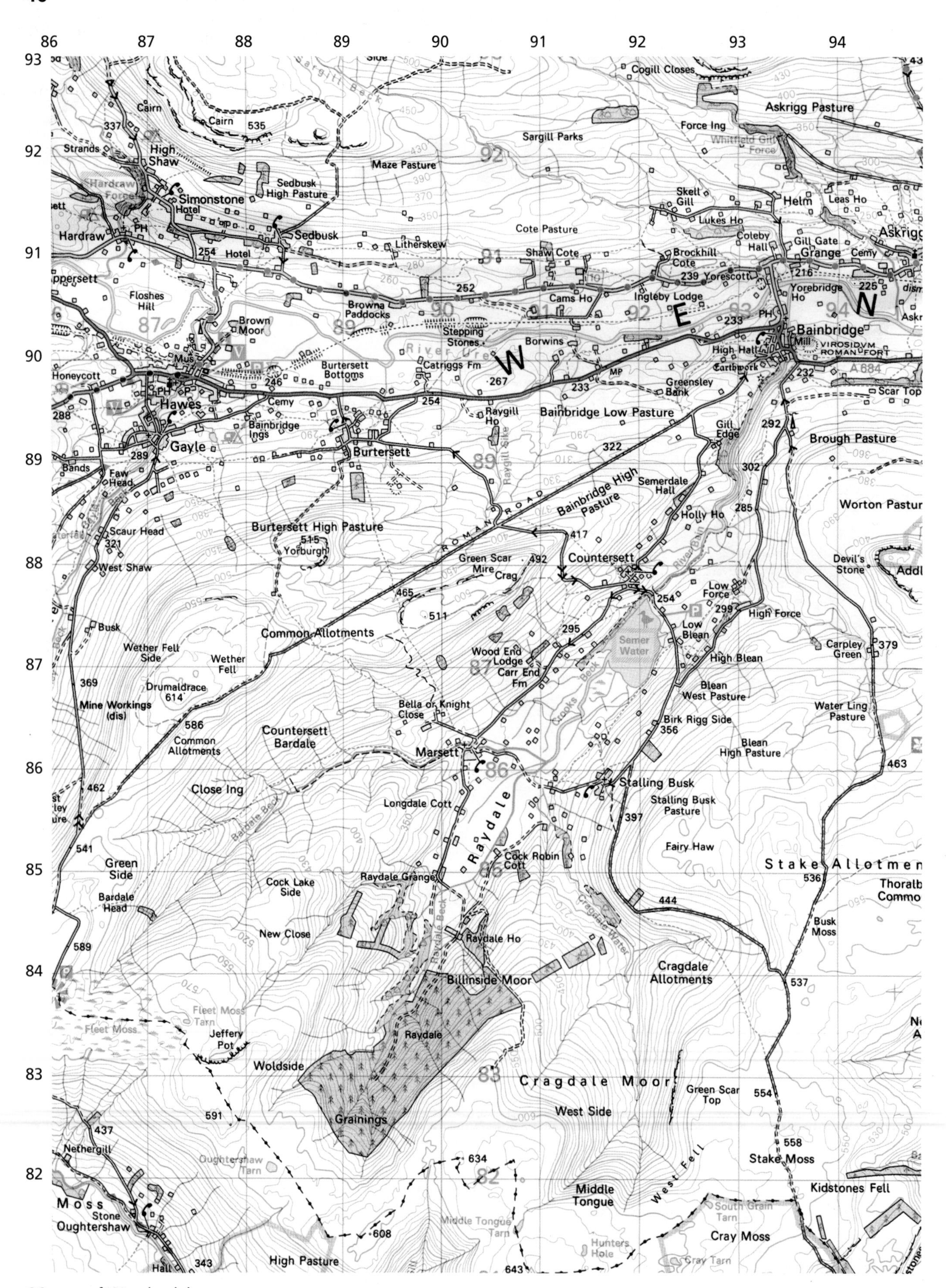

OS map of Wensleydale

3 Graphical Skills

In **Chapter 3 Graphical Skills** you will learn how to study:

- bar charts, histograms, compound bar charts and pyramid charts
- line graphs, compound line graphs, flow lines and isolines
- pie diagrams
- pictograms
- rose/ray diagrams
- triangular graphs
- topological diagrams and maps
- choropleth maps
- dispersion graphs
- proportional symbols
- scatter graphs.

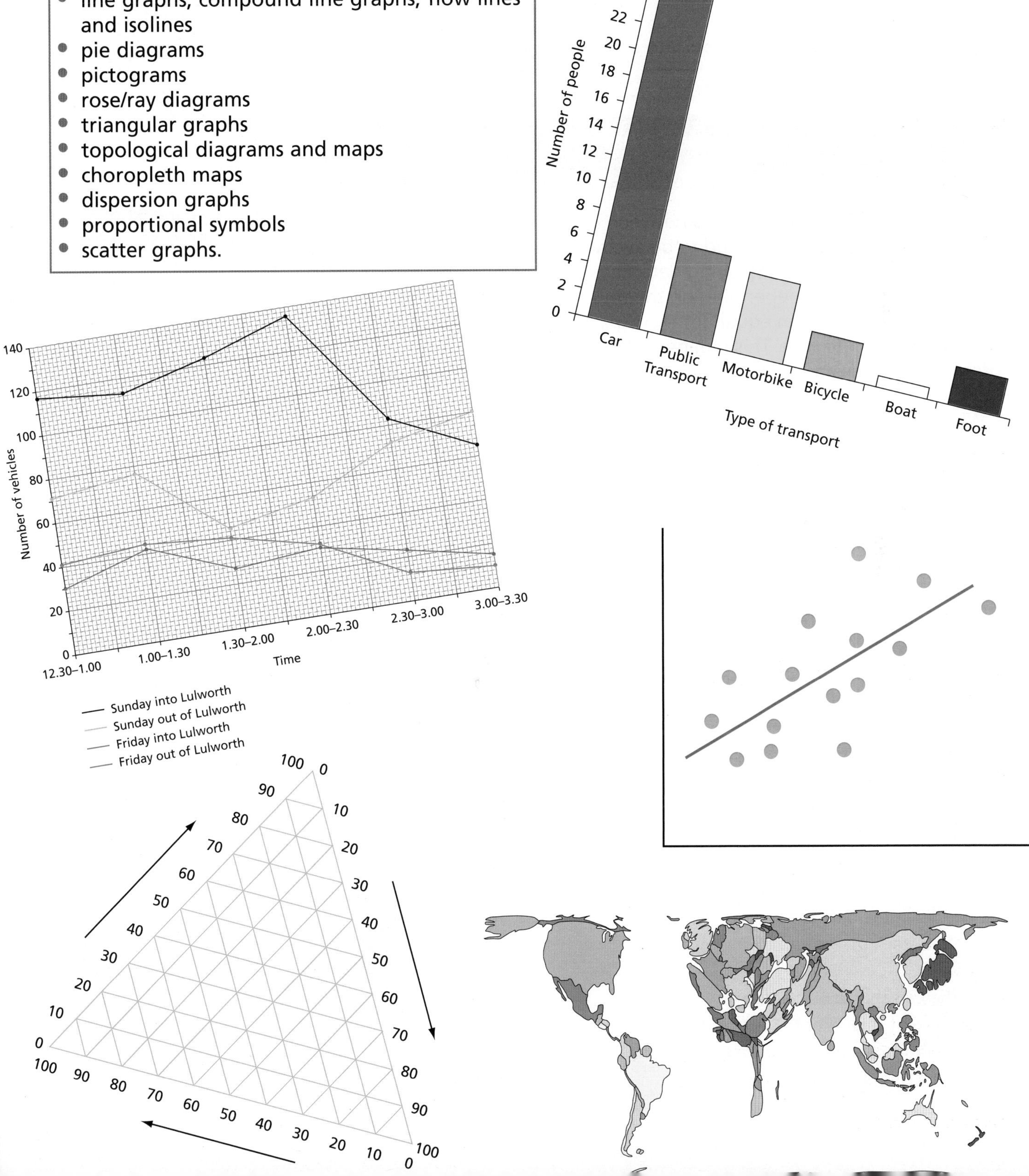

Different types of bar graphs

Learning objective – to study bar charts, histograms, compound bar charts and pyramid charts.

Learning outcomes

- To be able to construct each type of graph.
- To be able to describe and explain the patterns that the graphs show.
- To be able to suggest appropriate uses of these types of graphs.

What is the difference between a bar chart and a histogram?

Remember bar charts show distinct data; histograms show continuous data.

There are a number of differences between bar charts and histograms which students need to be aware of. Bar charts are one of the simplest forms of displaying data. Each bar is the same width but of varying length, depending on the figure being plotted. The bars should be drawn an equal distance apart. The data used for a bar chart is discrete data. This means that it is a distinct piece of information. An example would be the different types of vehicles in a traffic count.

When drawing a histogram, the bars should be drawn touching each other because a histogram is used to portray continuous data. An example would be the traffic flow for a continuous time frame.

How to draw a bar chart for the different types of transport used to get to a destination

- Decide on an appropriate scale on the *x*-axis for the bars. Remember the bars should be the same width with a gap in between each one which can be the same width as the data bar or a different width but the width must be a regular size.
- Decide on an appropriate scale on the *y*-axis for the number of vehicles. Remember the scale should be spaced out evenly and allow for the highest value in the data set.
- Draw each of the bars to the correct value.
- The bars should be coloured in different colours as the data is discrete.
- Don't forget to label your axes and put a title on your graph.

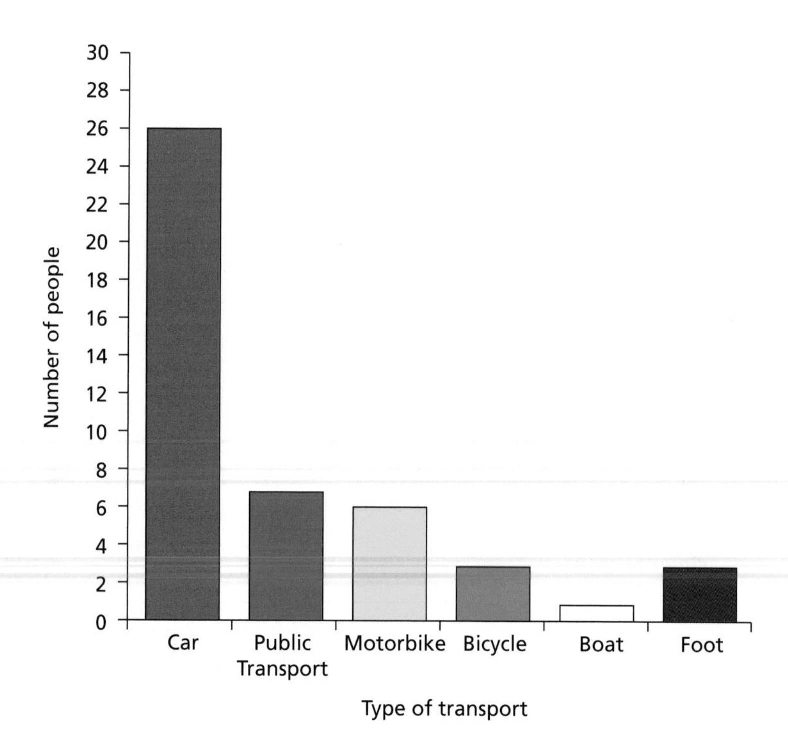

Figure 1 Bar chart showing the different types of transport used to get to a destination

How to draw a histogram of pedestrian flow for a continuous time scale

- Decide on an appropriate scale on the *x*-axis for the bars. Remember the bars should be the same width with no gap in between them.
- Decide on an appropriate scale on the *y*-axis for the number of people. Remember the scale should be spaced out evenly and allow for the highest value in the data set.
- Draw each of the bars to the correct value.
- The bars should be coloured in the same colour as the data is continuous.
- Don't forget to label your axes and put a title on your graph.

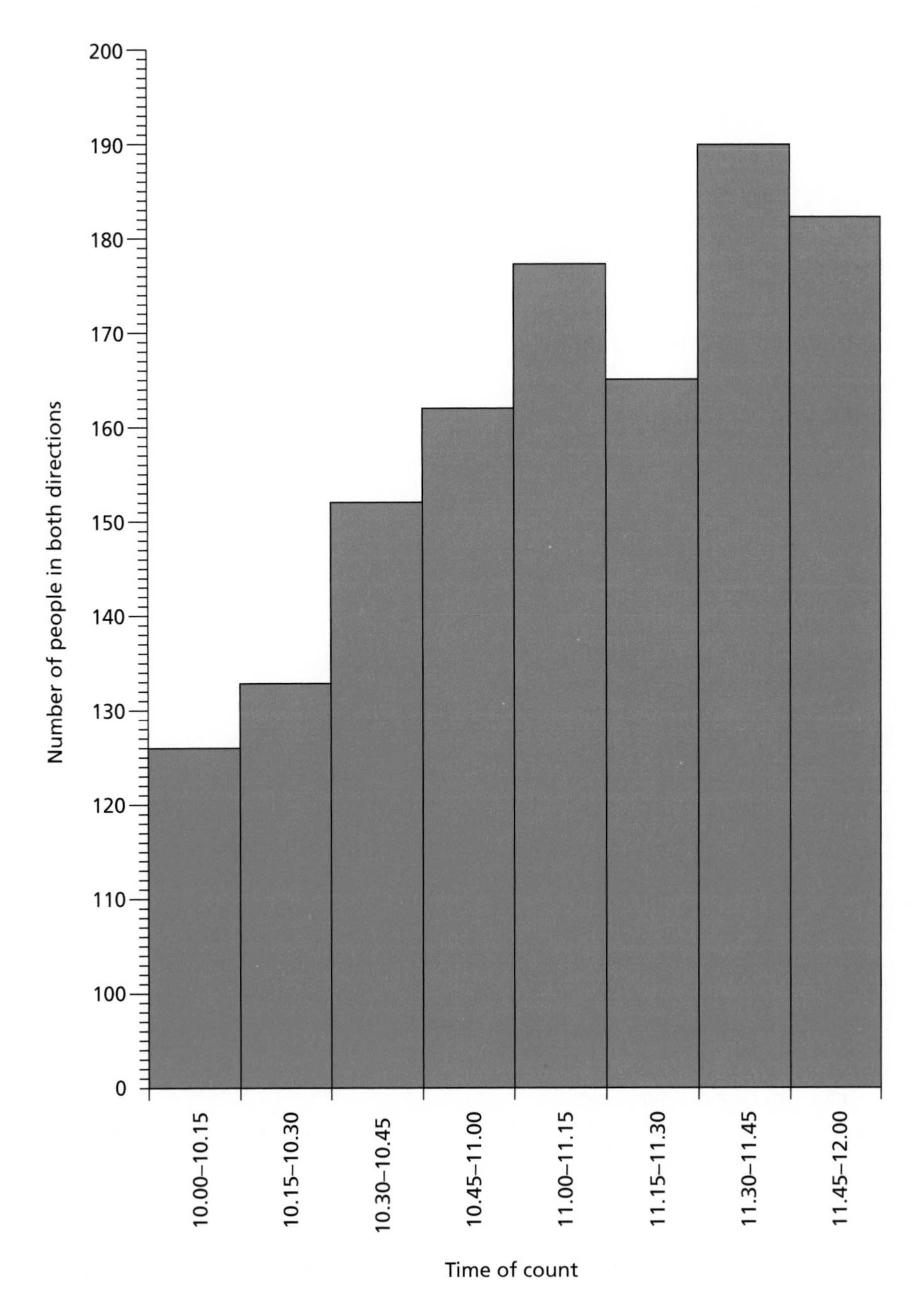

Figure 2 Histogram showing pedestrian flow for a continuous time scale

What is a compound bar chart?

In compound bar charts, the bars are subdivided on the basis of the information being displayed – in this example the sectors of industry in different countries. Each bar is out of 100 per cent.

How to draw a compound bar chart for sectors of industry in certain countries

- These can be drawn with the bars joined or an equal distance apart. In this example the bars are joined together.
- Decide on an appropriate scale on the *x*-axis for the per cent employed. Remember the scale should be spaced out evenly.
- Decide on an appropriate width of bar for the *y*-axis. Remember the bars should be the same width with no gap in between them.
- Divide each of the bars into the correct percentage for the industrial sector. For example, France has 6 per cent in primary, the line will be drawn at 6 per cent on the graph; 28 per cent in secondary, the line will be drawn at 34 per cent because it builds on the 6 per cent which is already there; tertiary has 66 per cent which is the rest of the space.
- Each sector, for example primary, secondary or tertiary, should be coloured the same colour in each of the bars.
- When complete the whole of the graph should be coloured in. In this example there will be three colours used.
- Don't forget to label your axes and put a title on your graph.

On the exam paper you may be asked to complete a bar graph:

- Use a ruler.
- Ensure the line or lines are drawn accurately.
- Complete the column by shading it exactly the same as the key.

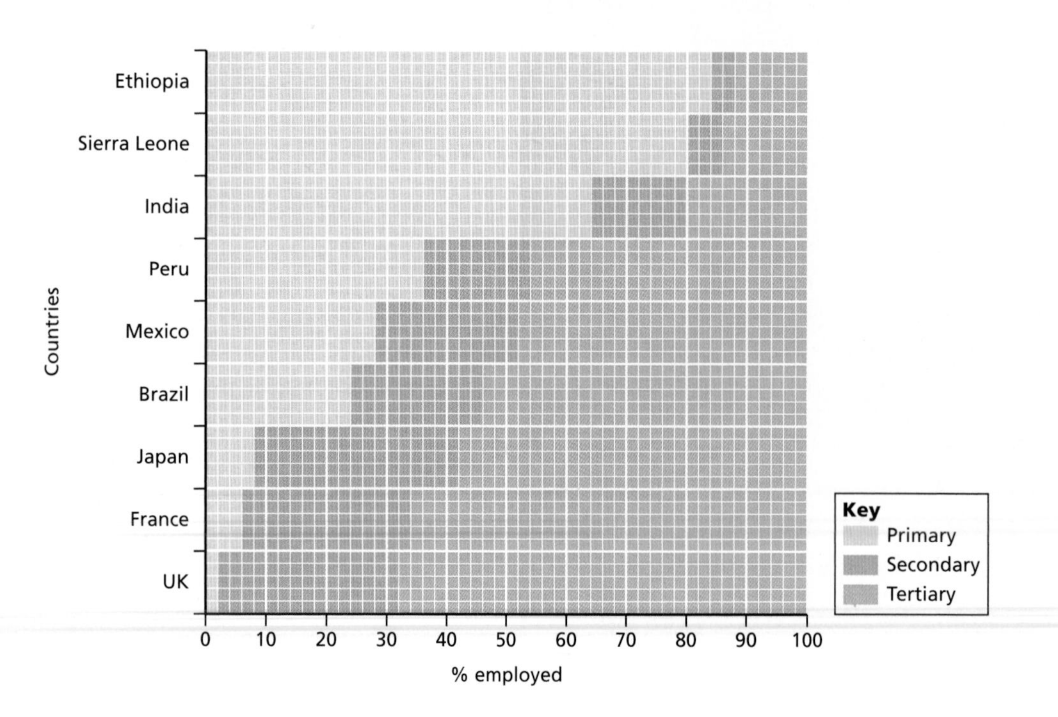

Figure 3 A compound or divided bar chart showing the sectors of industry in certain countries

What is a pyramid chart?

Another form of histogram is the pyramid graph. This is usually used to portray population (age–sex) data. It is constructed in a number of ways, but usually as 5- or 10-year age groups, with males on one side and females on the other. The lines are drawn horizontally and are the same width. The length of the bars is determined by the number of people in that age group or the percentage of the total population that age group represents. This type of chart can be used for any continuous data, for example a pedestrian count for a continuous time scale for movement in two directions, see Figure 2.

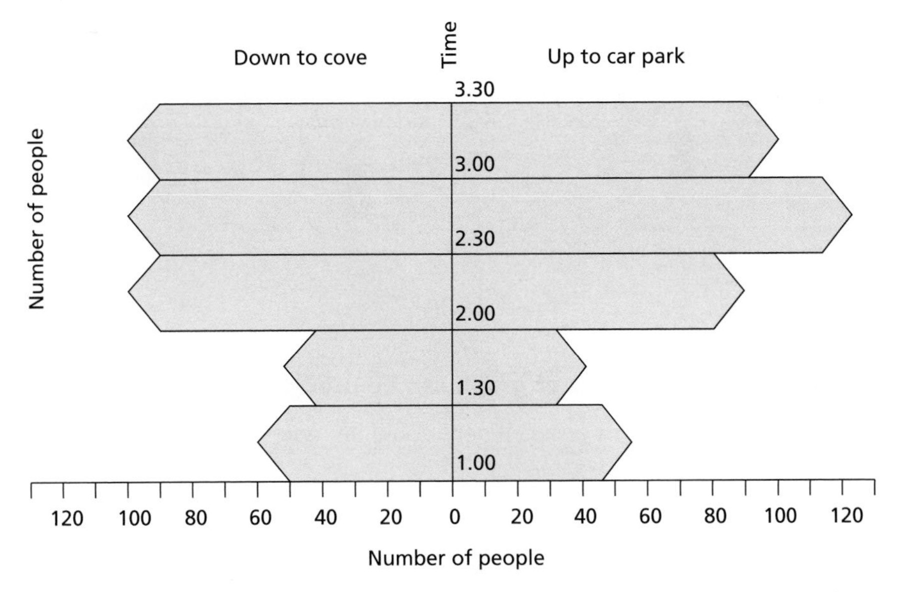

Figure 4 A pyramid chart of a pedestrian count completed outside the Gift Shop at Lulworth Cove in May 2009

How to draw a pyramid chart for a pedestrian count

- Use the graph paper in landscape view.
- Either draw two lines vertically up the centre of the paper a small distance apart, or draw one line as shown in Figure 4.
- The *x*-axis is now divided into two halves. The left-hand side of the graph paper is for the flow in one direction, the right-hand side is for the flow in the other direction. Decide on an appropriate scale for the *x*-axis. It should be the same for each side of the graph. Remember the scale should be spaced out evenly and allow for the highest value in the data set. Decide on an appropriate width of bar for the *y*-axis. Remember the bars should be the same width with no gap in between them.
- Draw each of the bars to the correct value.
- The bars should be coloured in the same colour as the data is continuous.
- Don't forget to label your axes and put a title on your graph.

What is a located bar chart?

Bar charts can be placed on to maps to give the data more relevance. This would be classed as a more sophisticated data presentation technique for your controlled assessment and is shown in Figure 5. In this case the bar charts have been drawn at the appropriate place on the map. Another way of doing this would be to draw the bar charts on tracing paper and use the tracing paper as an overlay. It is also possible to draw the bar charts on graph paper and paste them on to the map.

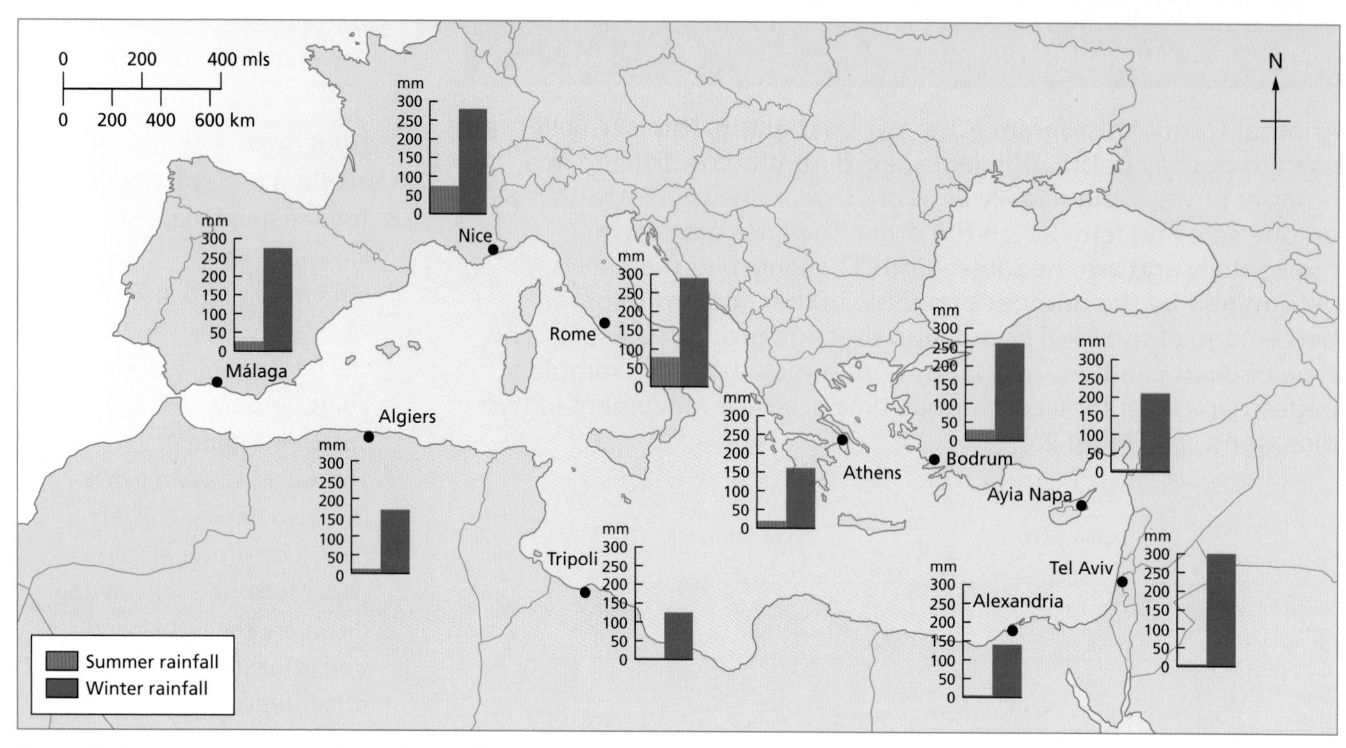

Figure 5 Seasonal water supply in the Mediterranean

Topics	Examples of where these types of graph can be used on the course
Challenges for the Planet	Percentage of carbon dioxide emissions per continent could be represented as a compound bar graph
Coastal Landscapes	The different distances of fetch could be portrayed as a histogram
River Landscapes	The number of global deaths from flooding over a 30-year period could be represented as a histogram
Glaciated Landscapes	Number of global deaths from avalanches over a 30-year period could be represented as a histogram
Tectonic Landscapes	Number of global deaths from tectonic hazards over a 30-year period could be represented as a histogram
A Wasteful World	Percentage of energy consumption per country could be represented as a compound bar graph
A Watery World	Changes in a country's use of water over time could be represented as a histogram
Economic Change	Change in number of people employed in primary industry could be represented as a histogram
Farming and the Countryside	Changes in the sale of organic produce over the past 20 years could be represented as a histogram
Settlement Change	The number of different services in a settlement could be portrayed as a bar graph
Population Change	The growth of world population could be represented as a histogram
A Moving World	Numbers of different types of short-term migrants could be portrayed as a bar graph
A Tourist's World	The increase in tourist numbers over a period of time could be portrayed as a histogram

ACTIVITIES

Higher tier

1 Construct a population pyramid for the data on Cambridge in the table below. Use the 'how to' box to help you.
2 Describe the shape of the pyramid. Use data in your answer.
3 Give reasons why this is an appropriate technique for this type of data.
4 State a graphical technique other than pyramids that could be used to display pedestrian flows.
5 Give a reason for your choice of technique in question 4.
6 Describe the pattern of seasonal water supply in the Mediterranean shown in Figure 5.

Foundation tier

1 Draw and complete the population pyramid for Cambridge shown in Figure 6. Use the data in the table below and the 'how to' box to help you.

Age	Males ×1000	Females ×1000
0–9	5	4
10–19	7	6
20–29	15	12
30–39	8	7
40–49	6	6
50–59	5	4
60–69	3.5	3
70–79	3	4
80–89	1	2.5
90 and over	1	2

2 Complete the sentences to describe the shape of the pyramid.

Up to the age of 69 there are more ...... in the population. In the 20–29 age group there are ...... more males than females. There are more ...... in the age groups 70 and above. In the 70–79 age group there is ...... more female than male.

3 Give reasons why this is an appropriate technique for this type of data. Choose the two most appropriate statements from the list below.
- It shows that there is a relationship between the data sets.
- It clearly shows if any of the age groups has more of the population than the others.
- The patterns in the data cannot be clearly seen.
- The data is discrete.
- The data is continuous.

4 Describe the pattern of seasonal water supply in the Mediterranean shown in Figure 5. Base your description on the answers to the questions below.
- What was the highest precipitation in the summer and which country received it?
- What was the highest precipitation in the winter and which country received it?
- Which country had the greatest difference between summer and winter totals?
- What was the amount?
- Which country had the smallest difference between summer and winter totals?
- What was the amount?

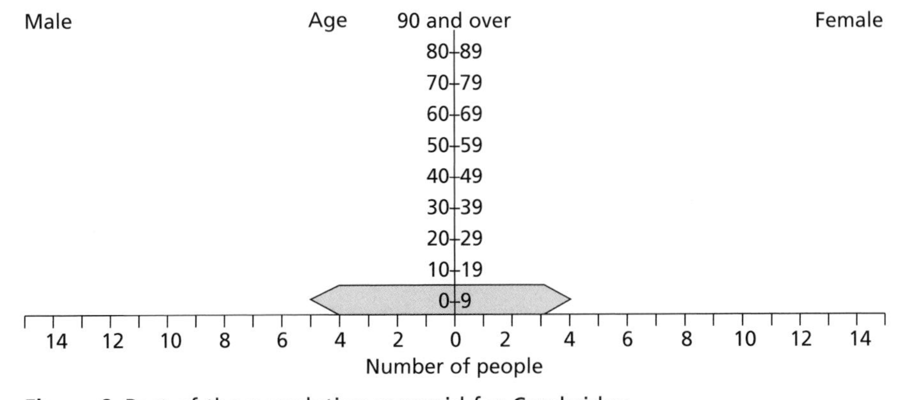

Figure 6 Part of the population pyramid for Cambridge

Review

By the end of this section you should be able to:
- construct bar charts, histograms, compound bar graphs and pyramid graphs
- describe and explain the patterns that the graphs show
- use the graphs appropriately.

STRETCH AND CHALLENGE

For the graphical technique you chose in higher tier question 4, use the data for pedestrian counts provided in Figure 4. Draw your suggested technique. Annotate it with the reasons as to why you think it is an appropriate technique.

Exam Tip

Remember on the exam paper you can be asked why a particular graphical technique is appropriate.

Line graphs, compound line graphs, flow line graphs and isolines

Learning objective – to study line graphs, compound line graphs, flow lines and isolines.

Learning outcomes
- To be able to construct each type of graph.
- To be able to describe and explain the patterns that the graphs show.
- To be able to suggest appropriate uses of these types of graphs.

What is a line graph?

A line graph is a basic graphical technique used to show changes over time (continuous data). In all line graphs there are independent and dependent variables. Line graphs can be used to show multiple sets of data, for example, traffic counts for the same place at different times.

How to draw a line graph for traffic counts

- The times of the count are the independent variable because the counts are dependent on when they are completed. These should be plotted on the horizontal *x*-axis.
- The number of vehicles is the dependent variable and should be plotted vertically on the *y*-axis.
- Decide on a scale appropriate for the range in values to be plotted on the *x*- and *y*-axes.
- Plot each line in turn. There should be four lines on your completed graph. They should be drawn in different colours and a key provided.
- Don't forget to label your axes and put a title on your graph.

	Friday 6 May		Sunday 8 May	
Times	**Into Lulworth**	**Out of Lulworth**	**Into Lulworth**	**Out of Lulworth**
12.30–1.00	30	41	117	71
1.00–1.30	42	44	113	76
1.30–2.00	27	41	123	45
2.00–2.30	30	32	136	53
2.30–3.00	23	13	83	72
3.00–3.30	15	10	65	80

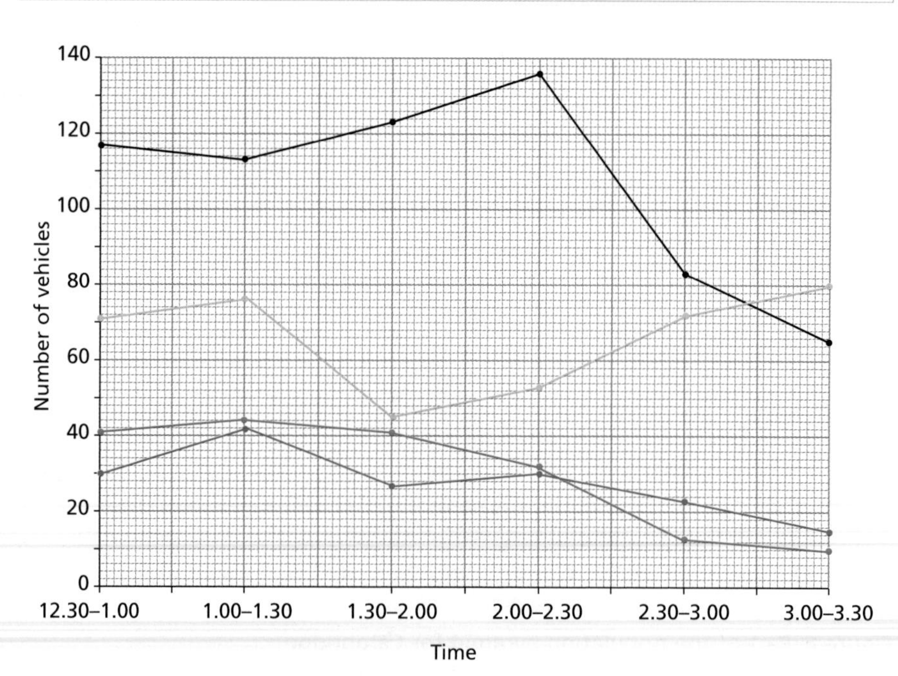

Figure 7 Graph of traffic flows into and out of Lulworth Cove

What is a compound line graph?

In compound line graphs, the graph is subdivided on the basis of the information being displayed in lines across the graph. For example, in Figure 8 the USA's percentage of world usage of renewable fuels has declined from 29 per cent in 1990 to a projected 20 per cent in 2030.

How to draw a compound line graph for renewable energy usage

- The years are the independent variable and should be plotted on the horizontal *x*-axis.
- The percentages are the dependent variable and should be plotted vertically on the *y*-axis.
- Decide on a scale appropriate for the range in values to be plotted on the *x*- and *y*-axes.
- Plot the data for the first country, the USA, on the graph.
- Add the percentage for the second country to the percentage for first country. This should calculated for each year, for example, USA in 1990 is 29 per cent; add 14 per cent for Canada. The point for Canada should be plotted at 43 per cent.
- The area between the countries should be coloured in a different colour for each section. The order of the countries should be largest to smallest with colours that can be clearly indentified.
- Don't forget to label your axes and put a title on your graph.

Remember you may be asked why a particular graphical technique is appropriate.

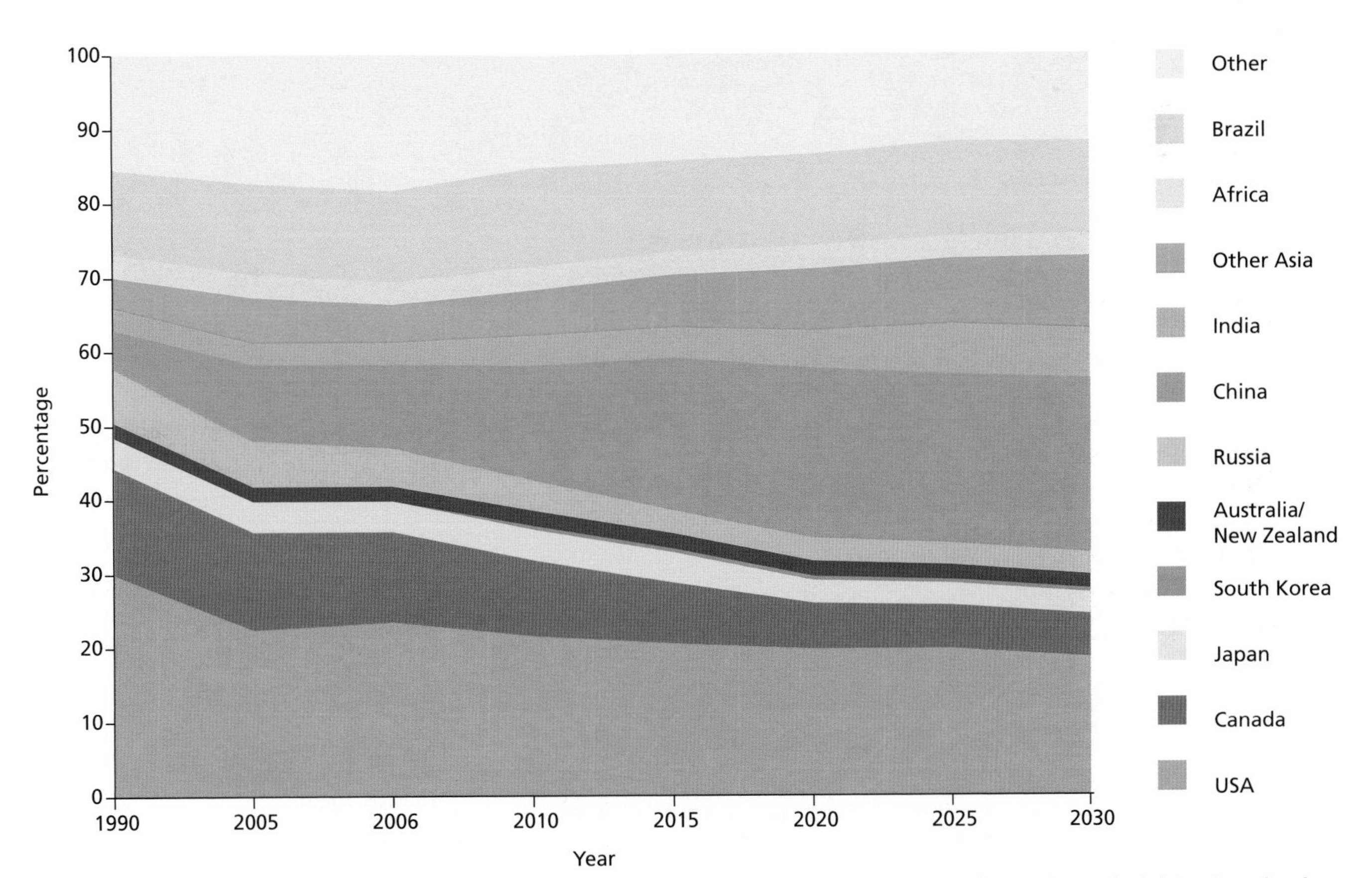

Figure 8 World usage of renewable energy by region 1990–2030 (Source: US Energy Information Administration data)

What is the difference between a flow line and an isoline?

Flow line maps and isoline maps are more sophisticated ways of portraying data.

A flow line shows movement between places. The thickness of the line indicates the amount of movement. The direction can be shown by an arrow. Figure 9 shows pedestrian flows at a number of places around Windsor town centre. The arrows point in both directions showing that the total is a combination of both directions.

How to draw a flow line map for pedestrian counts in Windsor

- Draw a base map which shows the relevant details, such as the main roads in the town centre where the counts were carried out. It would be a good idea at this point to mark in pencil where the actual counts took place.
- Study the range of the values and decide on an appropriate scale for the width of the arrows. If the scale is too large the flow lines will dominate the map so care must be taken at this stage to use the appropriate scale. In Figure 9 the scale is 5 mm thickness = 100 pedestrians.
- Draw the flow lines; these should go along the road where the count took place. The tail of the arrow should be where the flow began and the nose should point in the direction of the flow.
- As the count is a combination of both directions the arrow will point in both directions. If a note had been made of the direction of the count, two arrows should then be drawn, one for each direction.

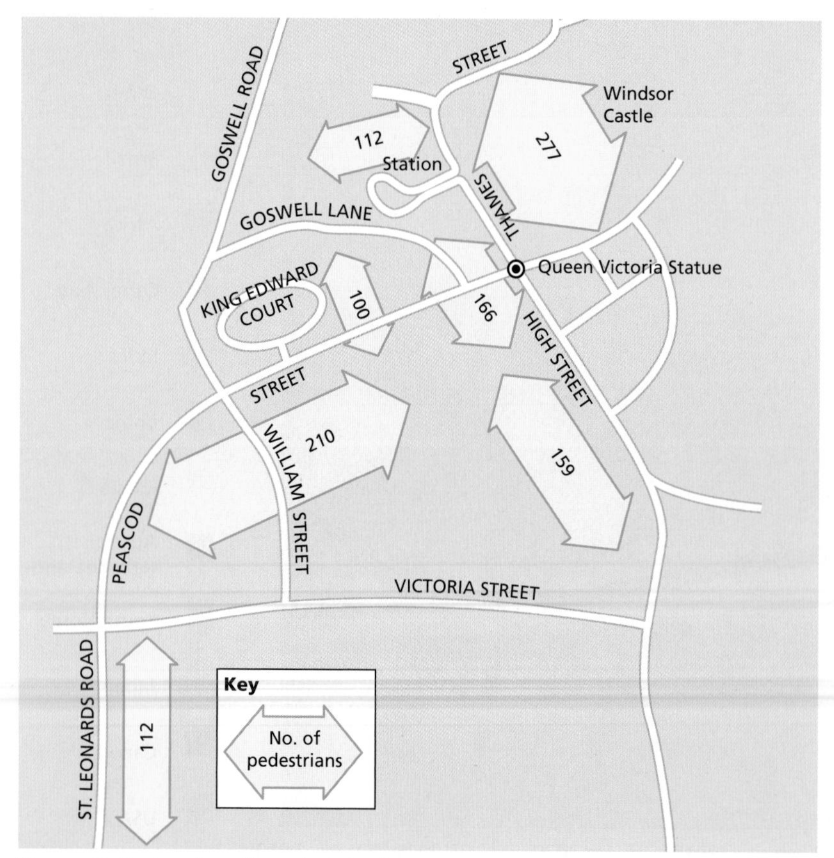

Figure 9 Pedestrian flows in Windsor

Street	Pedestrian flow
William Street	210
St. Leonards Road	112
High Street	159
Peascod Street (close to Queen Victoria Statue)	166
Peascod Street (close to King Edward Court)	100
Station	112
Thames Street	277

An isoline joins places of equal value and shows the distribution of an item over a particular area. Figure 10 shows places that are equal time from a shopping centre; these are known as isochrones. The most well-known isolines are:

- contours which join places on a map of equal height
- isobars which join places of equal pressure on a weather map
- isovels which join places of equal velocity in a river.

How to draw an isoline map for time distance from a retail centre

- Plot the data on to a map of the area. This should be a series of points which have a fairly even spatial distribution.
- Decide on the interval that you want between the isolines. If this is too small then the map will appear cluttered. If it is too great the map will become too generalised.
- Draw in your isolines, in this case every 5 minutes.
- The space between the isolines can be left blank or shaded in. If shaded the colour should become greater as the value becomes higher.
- If you shade your map then you should provide a key.

In the exam you would only be asked to complete isoline maps. Remember to draw lines evenly across the map.

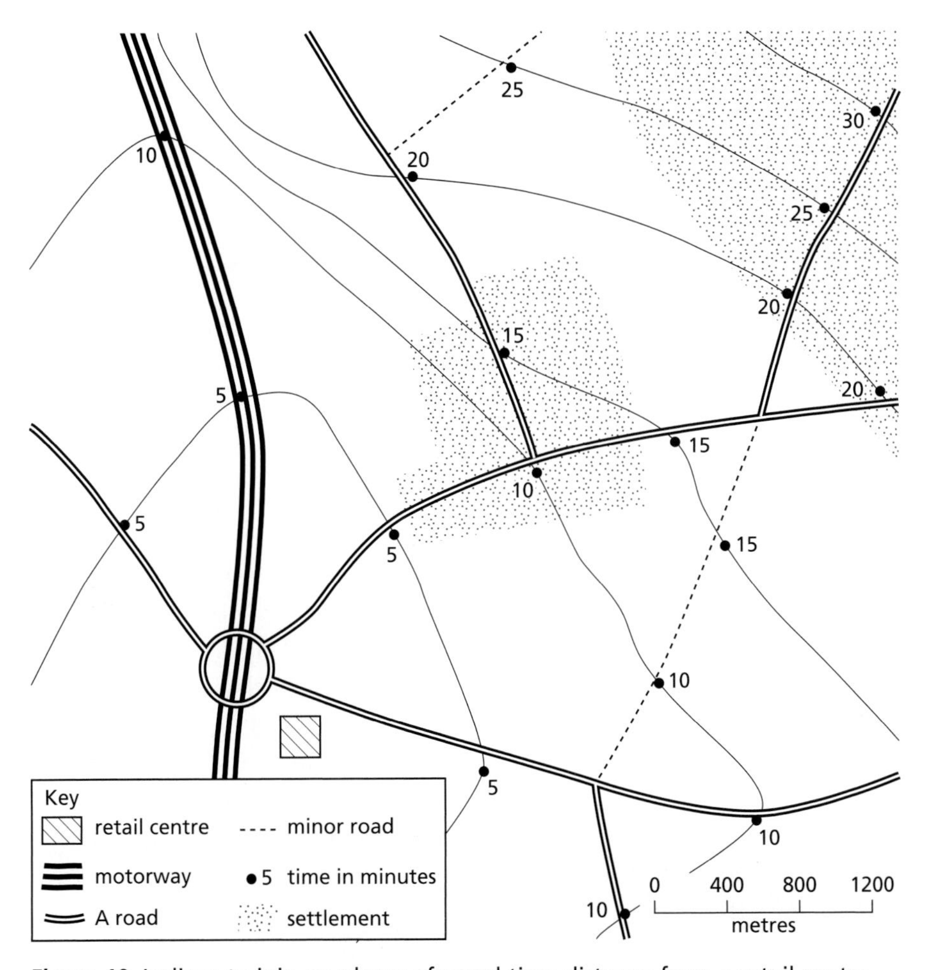

Figure 10 Isolines to join up places of equal time distance from a retail centre

Topics	Examples of where these types of graphs can be used on the course
Challenges for the Planet	Percentage of carbon dioxide emissions per continent could be represented as a compound line graph
Coastal Landscapes	Percentage change in wave height around the British coastline over a 20-year period time could be represented as an isoline map
River Landscapes	The number of global deaths from flooding over a 30-year period could be represented as a line graph
Glaciated Landscapes	The number of global deaths from avalanches over a 30-year period could be represented as a line graph
Tectonic Landscapes	The number of global deaths from tectonic hazards over a 30-year period could be represented as a line graph
A Wasteful World	The percentage of energy consumption per country could be represented as a compound line graph
A Watery World	Changes in a country's use of water over time could be represented as a line graph
Economic Change	Change in the number of people employed in a primary industry over time could be represented as a line graph. Different lines could represent different countries
Farming and the Countryside	Changes in the sale of organic produce over the past 20 years could be represented as a line graph
Settlement Change	The number of shoppers from different settlements could be represented as a flow line map
Population Change	The growth of world population could be represented as a line graph
A Moving World	The number of footballers from foreign countries who play for a particular club could be represented as a flow line mapgraph
A Tourist's World	Time–distance from a honeypot site could be represented as an isochrone map (an isoline map showing time differences)

Exam Tips

- You might be asked to complete line graphs or describe the changes that they show.
- Be sure to practise describing the changes not necessarily the patterns.
- You may be asked to describe one line on a graph that contains a number of lines – be sure to do what the question asks. For example, for a line graph that shows information on primary, secondary and tertiary employment, you could be asked just to discuss the changes in primary industry.

Review

By the end of this section you should be able to:

- construct line graphs, compound line graphs, flow lines and isolines
- describe and explain the patterns that the graphs show
- use the graphs appropriately.

ACTIVITIES

Higher tier

1 Study the line graph for traffic flows in and out of Lulworth Cove on page 54. Suggest two advantages and two disadvantages of this display technique.

2 A line graph is one way of portraying traffic count data. Choose another type of graphical technique:

   a Draw the graph.

   b Give reasons why this is an appropriate technique for this type of data.

   c Describe the pattern shown on the graph.

3 Study Figure 9 on page 56, which is a flow line map of pedestrian counts in Windsor. State four ways in which this map has been incorrectly drawn.

4 Redraw the map of pedestrian flows in Windsor correctly.

5 State another graphical technique that could be used to display this data.

6 Complete a compound line graph for the information in the table below. The table shows world share of gross domestic product (GDP).

	2005	2010	2015	2020	2025	2030
USA	23	22	21	21	20	18
Europe	28	24	22	18	16	13
Japan	10	9	8	6	5	4
China	4	9	12	14	16	18
India	3	4	5	8	10	12
Rest of Asia	6	6	7	8	8	10
South America	6	6	6	6	6	6
Russia	4	4	3	3	3	3
Other	16	16	16	16	16	16

Foundation tier

1 Look at the line graph for traffic flows into and out of Lulworth Cove on page 54. State one advantage and one disadvantage of using this technique for traffic data.

2 A line graph is one way of portraying traffic count data. Choose another type of graphical technique which can display this type of data.

   a Draw the graph.

   b Give reasons why this is an appropriate technique for this type of data.

   c Describe the pattern shown on the graph.

3 Look at Figure 9 on page 56, which is a flow line map of pedestrian counts in Windsor. It has been drawn incorrectly in a number of ways. State two.

4 Redraw Figure 9, taking account of all the errors.

5 State another graphical technique that could be used to display this data.

STRETCH AND CHALLENGE

Contour lines are isolines. Trace the drawing and complete it by plotting the following contour lines: 140, 130, 110, 100, 90, 80, and 70. Use page 57 to determine the type of slope shown on the diagram.

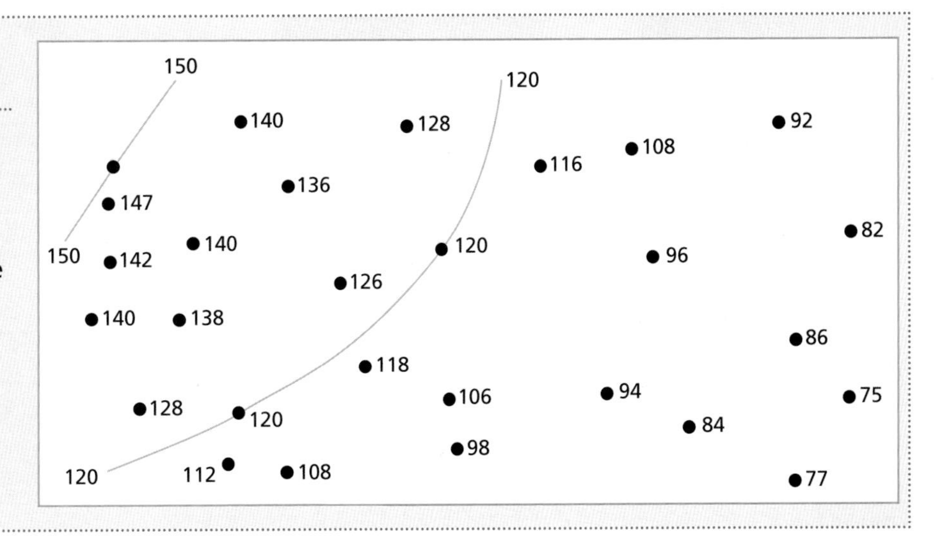

Pie diagrams

Learning objective – to study pie diagrams.

Learning outcomes
- To be able to construct pie diagrams.
- To be able to describe and explain the patterns that pie diagrams show.
- To be able to suggest appropriate uses of pie diagrams.

What is a pie diagram?

A pie diagram or divided circle is a basic graphical technique for showing a quantity which can be divided into parts. A pie diagram can also be drawn as a proportional circle. This is dealt with on pages 74–5 and would be classed as a sophisticated technique. Pie diagrams can be located on maps to show variations in the composition of a geographical phenomenon. They would then be regarded as a sophisticated presentation technique.

You will be required to construct, complete and interpret pie diagrams.

How to draw a pie diagram for types of employment in rural areas
- If necessary convert your data into percentages.
- You can either convert your percentages into degrees or use a percentage protractor to draw your pie diagram.
- Draw a circle and mark the centre.
- Subdivide the circles into sectors of the appropriate size.
- Differentiate the sectors by using different shadings or colours.
- Complete a key explaining the shadings and/or colours.
- Don't forget to put a title on your work.

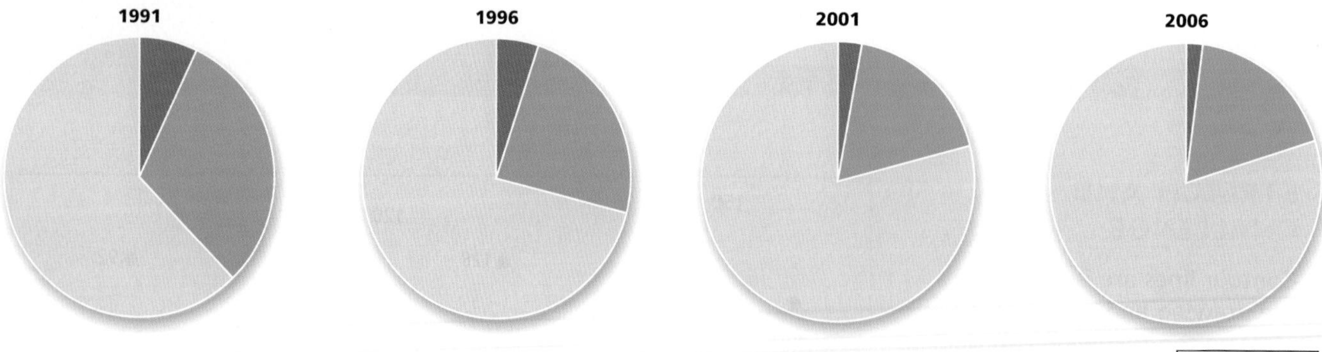

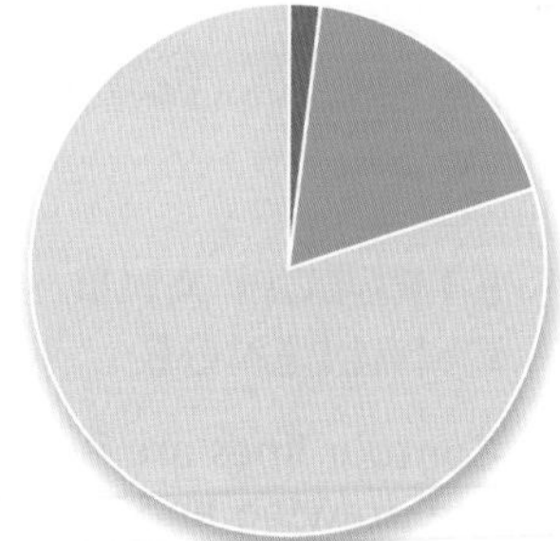

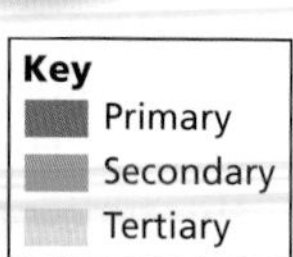

Figure 11 Employment in rural areas

Type of employment in 2006	Percentage
Primary	2
Secondary	18
Tertiary	80

ACTIVITIES

Higher tier

1 Construct a pie diagram for the 2006 data.
2 Outline the changes in employment shown between 1991 and 2006.
3 Suggest another graphical technique that could be used to show this information.
4 Justify your use of this technique.

Foundation tier

1 Construct a pie diagram for the 2006 data.
2 Complete the sentences to describe the changes in employment between 1991 and 2006.

In 1991 approximately ...... of the population in rural areas was employed in the primary sector. In 1991 ...... of the population was employed in the secondary sector. The numbers employed in the primary sector ...... between 1991 and 2006. The numbers employed in the tertiary sector ...... between 1991 and 2006.

3 Another technique that could be used for this data is a compound bar graph. Suggest one reason why this is an appropriate technique for changes in employment data.

Topics	Pie diagrams can be used to show
A Wasteful World	The different types of domestic waste produced by an HIC
A Watery World	The proportion of water used globally for domestic, agriculture and industry purposes
Economic Change	To portray the different sectors of industry in a country
Population Change	The number of people living in HICs and LICs
A Moving World	The number of migrants from different countries

STRETCH AND CHALLENGE

Complete a pie diagram for either the suggested watery or wasteful examples given below.

Type of water consumption	Percentage usage
Domestic	11
Agriculture	69
Industry	20

Type of country	Percentage consumption of world commodities
HIC	86
MIC	13
LIC	1

Review

By the end of this section you should be able to:

- construct pie diagrams
- describe and explain the patterns that the diagrams show
- use the graphs appropriately.

Pictograms

Learning objective – to study pictograms.

Learning outcomes
- To be able to construct pictograms.
- To be able to describe and explain the patterns that pictograms show.
- To be able to suggest appropriate uses of pictograms.

What is a pictogram?

A pictogram is a way of portraying data using appropriate symbols or diagrams which are drawn to scale.

How to draw a pictogram for world population growth

- Study the data that has to be displayed. Decide on the symbols to use to display the information. In this case the shape of a person could be drawn.
- Decide the scale for the symbols. In this case 1 person = 1 billion people.
- When constructing a pictogram, the years do not need to be continuous.
- Draw the symbols. In some cases the symbol may not be full sized if it represents a proportion of the full amount.
- Don't forget to provide a key and put a title on your graph.

Topics	Pictograms can be used to show
A Wasteful World	Carbon emissions of selected countries
A Watery World	The increasing number of people who own washing machines and dishwashers in the UK
Economic Change	The increase in machinery used on farms
Population Change	Hours of sunshine in Spain
A Moving World	Car ownership in the UK

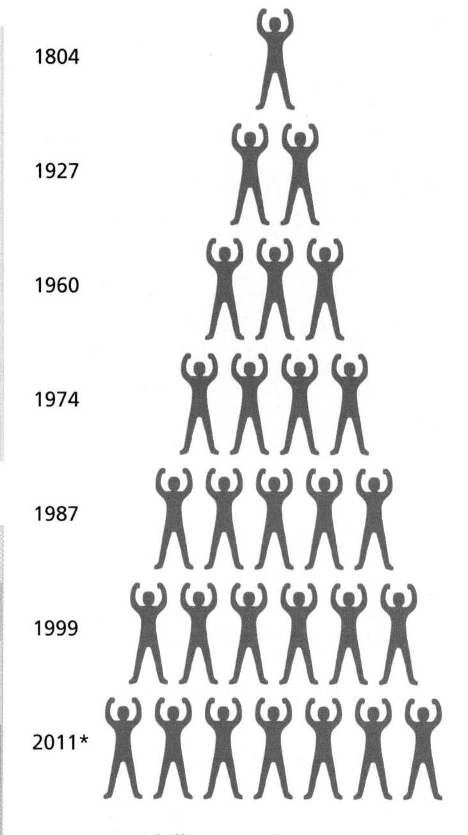

Figure 12 Pictogram for world population growth

ACTIVITIES

Higher tier

1 Construct a pictogram for the data below which was taken from a questionnaire of how shoppers travelled to a shopping centre.

Mode of transport	Number of people
Car	26
Public transport	7
Taxi	1
Pedal bike	8
Motorbike	3
On foot	5

2 Choose another graphical technique. Display the data using this technique.

3 Suggest two advantages of using a pictogram.

4 Suggest two advantages of using the other technique you have chosen.

Foundation tier

1 Use the 'how to' box to construct a pictogram for the data below. Hint: use a drawing of a car to represent the car total, etc. Draw one car to represent two people. You should use the same scale for each of the vehicles you draw.

Mode of transport	Number of people
Car	26
Public transport	7
Taxi	1
Pedal bike	8
Motorbike	3
On foot	5

2 This data could also be displayed in a bar chart. Draw a bar chart to display this information.

3 Compare the two techniques. They both have advantages for displaying this data. Choose the one you think is the best. Give two reasons to justify your choice.

STRETCH AND CHALLENGE

Pictograms can be located on maps to produce a more sophisticated technique. Use the data below to draw pictograms on the map of Windsor (refer to the map on page 56 for the location of the streets).

Street	Pedestrian flow
William Street	210
St. Leonards Road	112
High Street	159
Peascod Street (close to Queen Victoria Statue)	166
Peascod Street (close to King Edward Court)	100
Station	112
Thames Street	277

When drawing pictograms always remember to provide a key for the symbols.

Review

By the end of this section you should be able to:

- construct pictograms
- describe and explain the patterns that pictograms show
- use pictograms appropriately.

Rose diagrams and triangular graphs

Rose/ray diagrams

Learning objective – to study rose/ray diagrams.

Learning outcomes
- To be able to construct rose/ray diagrams.
- To be able to describe and explain the patterns that rose/ray diagrams show.
- To be able to suggest appropriate uses of rose/ray diagrams.

What is a rose or ray diagram?

A rose/ray diagram consists of straight lines which show movement or a connection between two places. These can be drawn on base maps to create a sophisticated technique. There are several types of ray diagram, including:

- desire lines that show movement from one place to another
- wind roses that have rays originating from the point at which they are being measured. The rays point to the wind direction. The length of the ray is drawn to scale for the amount of days the wind has blown from that particular direction.

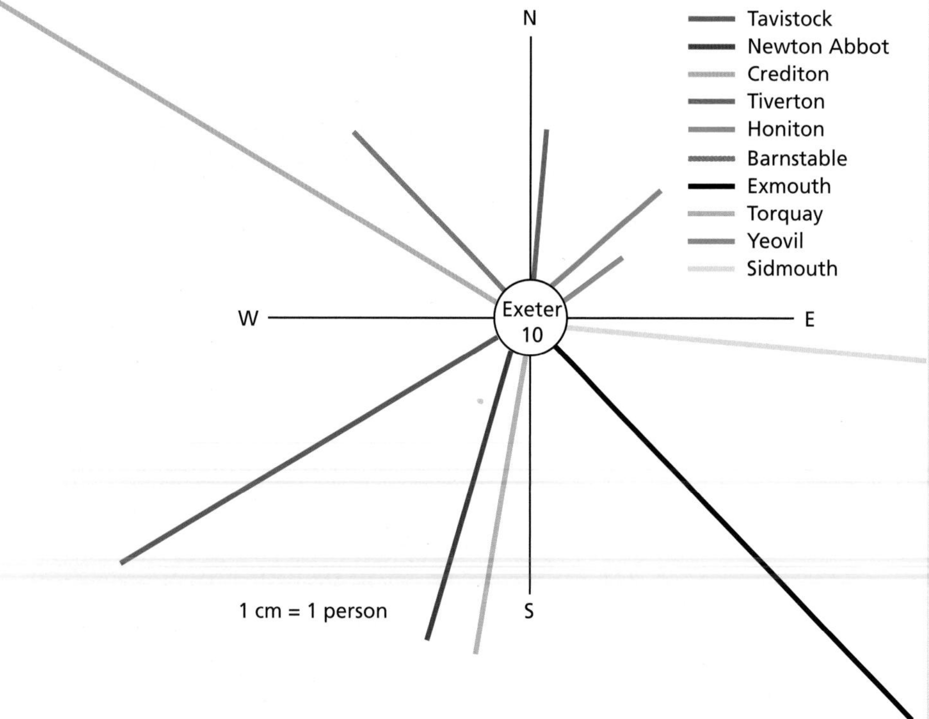

Figure 13 A rose/ray diagram for people who shop in Exeter

How to draw a rose/ray diagram (desire lines) for people who shop in Exeter
- Work out the number of people coming from each direction to Exeter.
- Work out a scale which is appropriate for the values in your data.
- Draw the rays of the diagram in the correct direction for the location of the place that the shoppers came from. This must be to scale. This can be done on paper or on a base map of the area concerned.
- Each ray should be completed in a different colour.
- A circle should be drawn for Exeter with the number 10 (the total number of rays) in the middle.
- Don't forget to provide a key and put a title on your graph.

Topics	Rose/ray diagrams can be used to show
A Moving World	The number of people coming to the UK from different European countries
A Tourist's World	The number of tourists coming to the UK from different countries

ACTIVITIES

Higher tier

1 Construct a ray diagram for the data below. Remember to draw the arrows in the correct direction to where the country is located; even though the information is not being put on to a map it should be geographically correct.

Top ten overseas tourist destinations from the UK	Number of British tourist visits (millions)
Spain	14.0
France	11.0
USA	4.0
Irish Republic	3.9
Italy	3.5
Germany	3.0
Portugal	2.4
Greece	2.1
Netherlands	2.0
Turkey	2.0

2 Describe the pattern shown by the ray diagram.

3 Suggest reasons why this is a good way of displaying this type of data.

4 Outline another technique that could be used to display this data.

5 Compare and contrast the two techniques in their ability to display this information.

Foundation tier

1 Copy out and complete the following ray diagram showing the top ten overseas tourist destinations for British holidaymakers. Use the data in the table opposite to complete the diagram.

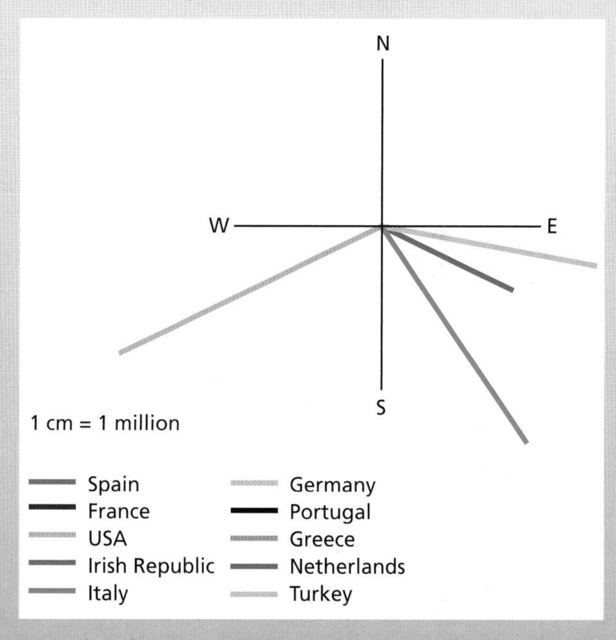

2 Describe the pattern shown on the ray diagram.

3 Another way of displaying this data would be in a histogram. Suggest reasons why a histogram is an appropriate graphical technique for this type of data.

4 Which do you think is the best technique for this data? Justify your choice.

Wind rose diagrams are not part of this specification and therefore will not appear on exam papers. Other rose and ray diagrams may appear.

STRETCH AND CHALLENGE

Use the data on the top ten tourist destinations. Complete a ray diagram on a suitable map.

Review

By the end of this section you should be able to:

- construct ray diagrams
- describe and explain the patterns that ray diagrams show
- use ray diagrams appropriately.

Triangular graphs

Learning objective – to study triangular graphs.

Learning outcomes

- To be able to construct triangular graphs.
- To be able to describe and explain the patterns that triangular graphs show.
- To be able to suggest appropriate uses of triangular graphs.

What is a triangular graph?

Triangular graphs have axes on three sides and so they are used to display data which can be divided into three. The data must be in percentages that total 100 per cent. The main advantage is that the graph displays a large amount of data on one chart. It is classed as a sophisticated data technique because it displays information about more than one variable.

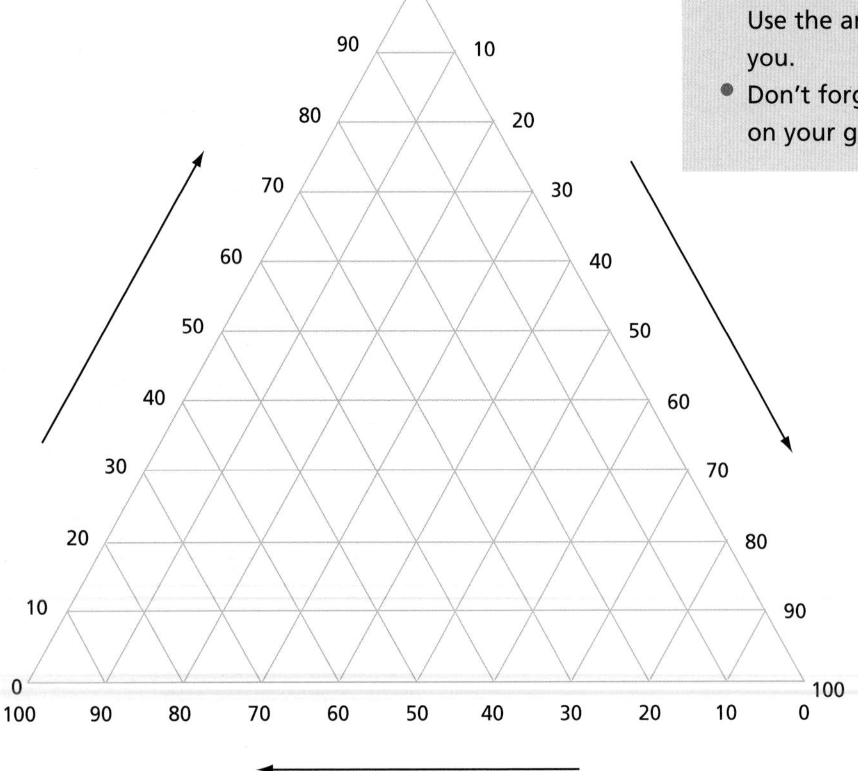

Figure 14 A base chart for triangular graphs

How to draw a triangular graph for types of employment in different countries

- There are three axes on the graph. Each axis represents 100 per cent marked at 10 per cent intervals.
- At each 10 per cent interval, lines are drawn at 60° to carry the values across the graph. On Figure 14 the axes should be labelled primary, secondary and tertiary.
- Plot the data for types of employment in different countries from the table below on to a copy of the triangular graph in Figure 14. Use the arrows around the graph to guide you.
- Don't forget to label your axes and put a title on your graph.

Country	Primary	Secondary (%)	Tertiary (%)
Ethiopia	84	4	12
Sierra Leone	80	6	14
India	64	16	20
Peru	36	18	46
Mexico	28	24	48
Brazil	24	22	54
Japan	8	34	58
France	6	28	66
UK	2	30	68

Topics	Triangular graphs can be used to show
A Watery World	Domestic, industrial and agricultural usage of water
A Wasteful World	Types of waste: domestic, industrial, educational

ACTIVITIES

Higher tier

1 Construct the triangular graph which is explained in the 'how to' box. A base graph has been provided in Figure 14.
2 Describe the pattern shown in the triangular graph you have constructed.
3 Outline the advantages and disadvantages of triangular graphs.
4 How many pie charts would have to be drawn to portray the same amount of data?

Foundation tier

1 Construct the triangular graph as explained in the 'how to' box.
2 Complete the following sentences to describe the pattern shown by the triangular graph.
The country with the highest number of people employed in the primary sector is ...... with 84 per cent. The country with the lowest number of people employed in primary industry is the UK with ...... The country that is spread most evenly between the sectors is ...... The country with the largest number employed in secondary is ...... France has the second highest tertiary sector with ...... employed.
3 Triangular graphs have many advantages and disadvantages. Look at the following statements. Draw up a table headed advantages and disadvantages and put the statements into the correct columns.
- They are very difficult to construct.
- They are able to show three sets of data on one graph.
- They can be confusing if you are unsure which way to read the lines.
- They have to be constructed for data which adds up to 100 per cent.
- It is very easy to see the patterns they portray.
- Anomalies are easy to spot.
- Base triangles have to be provided because they are very difficult to construct.
- They can save a lot of time and effort because a lot of information is on one graph.

In the exam you will not have to draw out a complete triangular graph, but you may have to complete one. Hint: there will always be guidance arrows to help you.

STRETCH AND CHALLENGE

Plot the data in the table on page 66 as pie charts on a map of the world. This is a sophisticated data presentation technique.

Review

By the end of this section you should be able to:
- construct triangular graphs
- describe and explain the patterns that triangular graphs show
- use triangular graphs appropriately.

Topological diagrams

Learning objective – to study topological diagrams and maps.

Learning outcomes
- To be able to construct topological diagrams and maps.
- To be able to describe and explain the patterns shown by topological diagrams and maps.
- To be able to suggest appropriate uses of topological diagrams and maps.

What is a topological diagram?

In the exam you would only be asked to complete a topological map. You may be asked why a topological map has been drawn to display that particular type of data.

There are two types of topological diagram. The first type includes diagrams or maps of routes where the position of the place remains the same but the actual distance and direction are not so important. The other type of topological diagram or map shows the areas of a place but the size is determined by the value being portrayed.

- Maps of route networks. This is a map that has been made simpler. The actual distance and direction are not so important, but the position of places remains the same. If you were asked to give directions to somewhere you would do it by perhaps drawing a map. The map would include the important features that you would pass on the route, such as large roads, railway lines and rivers. A good example is the London Underground map.
- Maps of areas such as countries where the size of the country has been changed. The size of the country is no longer its geographical size because it has been drawn to represent a different value. An example of this would be the number of migrants to a country.

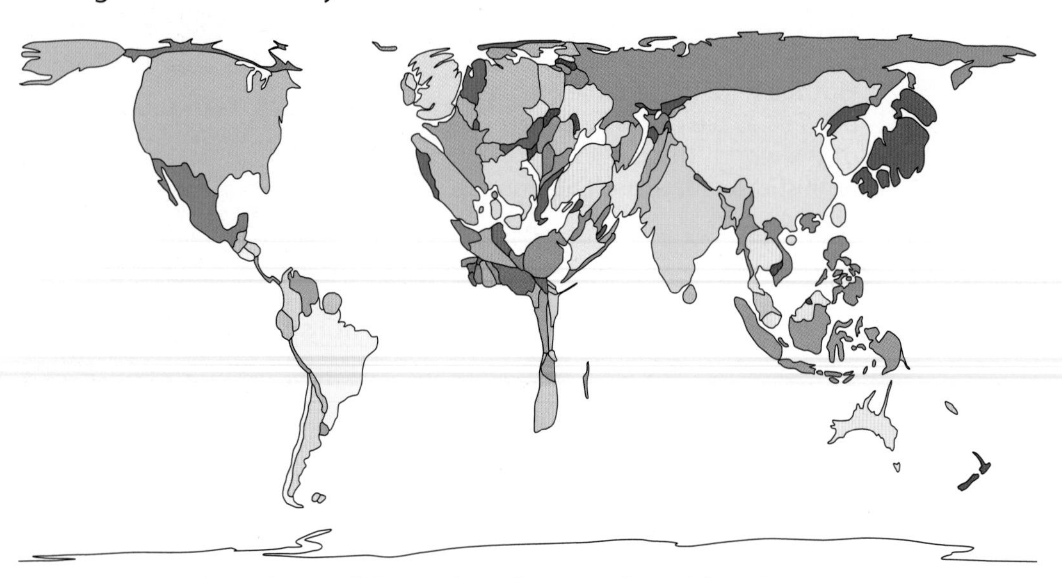

Figure 15 A topological map of the amount of waste collected from homes, schools and businesses

How to draw a topological map of your route to school

- On a piece of paper put the location of the house where you live and the location of your school. These should be in their correct situation in relation to the position of the places.
- Draw the roads that you would take from your house to school.
- Try to add some of the road names.
- Add on any railway lines and rivers that you would pass or cross on the route using different colours.
- Local services such as churches and public houses could also be marked.
- Add a key.

ACTIVITIES

Higher tier

1 Complete a topological map for your journey to school.
2 What are the advantages of topological maps?
3 Download and print a copy of the London Underground map (www.tfl.gov.uk/assets/downloads/standard-tube-map.pdf) and outline how you would travel on the Underground between the following places. You should describe two different routes:
 a Waterloo station to Euston station.
 b Victoria station to Liverpool Street station.
4 Using topological maps has a number of disadvantages. State two disadvantages of using the London Underground map.

Foundation tier

1 Complete a topological map for your journey to school.
2 Download and print a copy of the London Underground map (www.tfl.gov.uk/assets/downloads/standard-tube-map.pdf) to outline how you would travel on the underground between the following places:
 a Paddington station to Liverpool Street station.
 b Paddington station to Waterloo station.
3 State one advantage and one disadvantage of using topological maps.

You might be asked to complete a topological map or assess their usefulness compared to other types of maps.

STRETCH AND CHALLENGE

Study Figure 15, a topological map of the amount of waste collected from homes, schools and offices. Outline the main features of the map.

Review

By the end of this section you should be able to:

- construct topological diagrams
- describe and explain the patterns that the topological diagrams show
- suggest appropriate uses of topological diagrams.

Choropleth maps

Learning objective – to study choropleth maps.

Learning outcomes

- To be able to draw choropleth maps.
- To be able to describe and explain the patterns that choropleth maps show.
- To be able to suggest appropriate uses of choropleth maps.

What is a choropleth map?

A choropleth map is a map that is shaded according to a prearranged key, each shading or colour representing a range of values. The colours should become darker as the numbers increase. There are some inaccuracies in using this technique; one of these is that variations within units are concealed. It also gives a false impression of abrupt changes at boundaries. However, choropleth maps are easy to complete and show a good visual impression of change over space.

Topics	Choropleth maps can be used to show
A Wasteful World	Waste generated by different countries of the world
A Watery World	World water scarcity
A Moving World	Showing where the footballers who play at a British club come from, for example Arsenal
A Tourist's World	The top ten source countries for tourists to the UK

How to draw a choropleth map for average precipitation in the UK (see Figure 16)

- Obtain a base map of the UK.
- Find the range of your values and devise a shading scale.
- You should try to have no fewer than four shading bands and no more than eight.
- The shading should get darker as the value gets higher.
- Complete the map by shading in the areas.
- Draw a key for the map.
- Don't forget to put a title on your map.

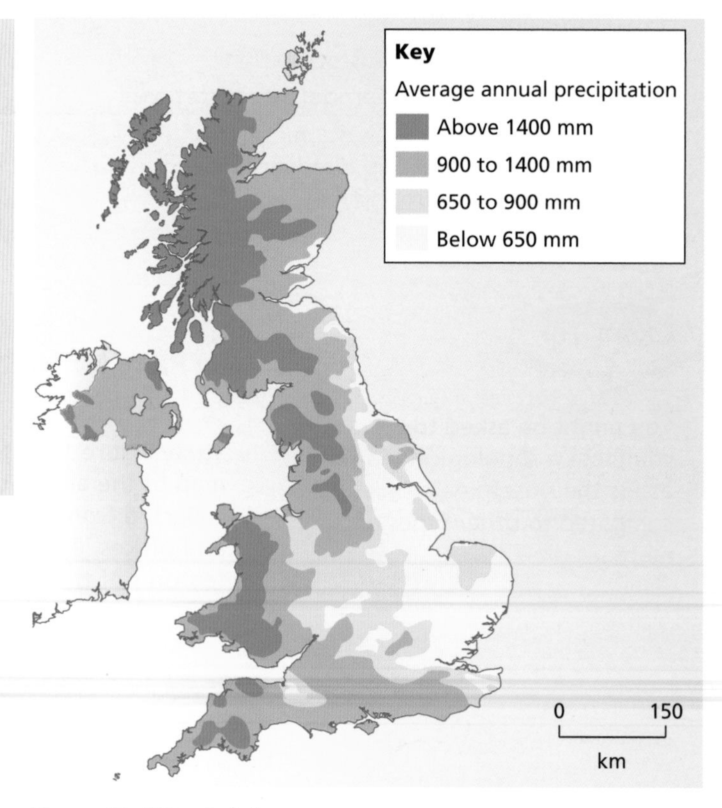

Figure 16 UK rainfall

ACTIVITIES

Higher tier

1. Draw a choropleth map for the information on the top ten tourist destinations in 2009 which is shown in the table on the right.
2. A choropleth map is one way that this data could be displayed. Suggest two advantages and two disadvantages of using a choropleth map for this data.
3. Suggest and complete another sophisticated data presentation technique that could have been used to display the data.
4. State one way that this technique is more appropriate than a choropleth map and one way that it is less appropriate than a choropleth map.
5. A basic technique for displaying this data would be a histogram. Why would a histogram be drawn rather than a bar chart?

Foundation tier

1. Draw a choropleth map for the information on the top ten tourist destinations in 2009 which is shown in the table below.

Destination	Tourist arrivals (millions)
France	81.9
Spain	59.2
USA	56.0
China	54.7
Italy	43.7
UK	30.7
Germany	24.4
Ukraine	23.1
Turkey	22.2
Mexico	21.4

2. Describe the pattern shown by the map.
3. Suggest two reasons why a choropleth map is a good way of displaying this data.
4. Another sophisticated technique that could be used to display this type of data would be flow lines. Complete a flow line map for the data on the top ten tourist destinations in 2009.
5. A basic technique for displaying this data would be a histogram. Why would a histogram be drawn rather than a bar chart?

STRETCH AND CHALLENGE

Use the data in the table below to draw a choropleth map of the source countries of migrants to the UK from Eastern European countries between 2004 and 2006.

Country	Number of migrants (thousands)
Czech Republic	25
Estonia	10
Hungary	20
Latvia	28
Lithuania	55
Poland	260
Slovakia	40
Slovenia	5

Questions could be set on completing a choropleth map and a key. Therefore, be sure you understand the rules for a successful choropleth map.

Review

By the end of this section you should be able to:

- draw choropleth maps
- describe and explain the patterns that choropleth maps show
- use choropleth maps appropriately.

Dispersion graphs

Learning objective – to study dispersion graphs.

Learning outcomes

- To be able to construct dispersion graphs.
  To be able to describe and explain the patterns that dispersion graphs show.
- To be able to suggest appropriate uses of dispersion graphs.

What is a dispersion graph?

A dispersion graph shows the range of a set of data. It shows if the data tends to group or disperse. It can also be used to compare sets of data. The values are plotted on the vertical axis. There is also a short horizontal axis which can show the frequency (the number of times) the variable occurs.

How to draw a dispersion graph for the size of pebbles found on a beach

- The study was completed at three sites, therefore the graph must allow for this.
- Study the values of the pebble sizes in the sample. Decide on a scale for the range of values.
- Plot the scale on the vertical axis.
- The horizontal scale is for the sites and if necessary allows you to repeat a value.
- Don't forget to label your axes and put a title on your graph.
- The mean size has also been marked on Figure 17.

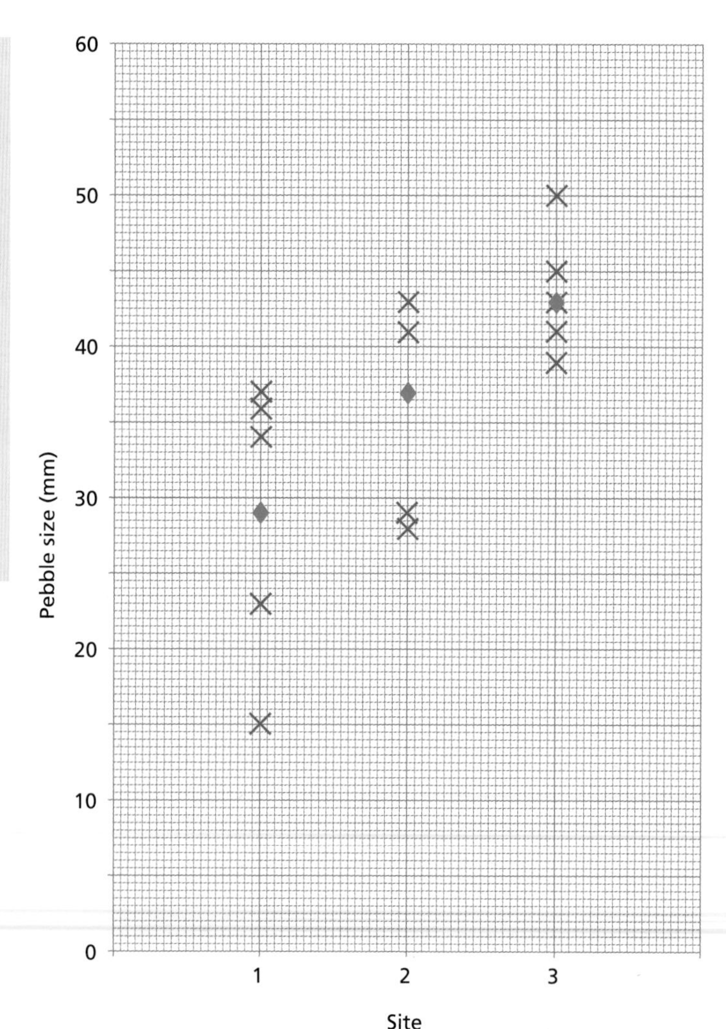

Figure 17 Pebble sizes at three sites on Walton beach

ACTIVITIES

Higher tier

1 Complete a dispersion graph for the data in the tables below. The tables show data on pebble sizes gathered by a group of students from two sites on the Afon Tarell in South Wales.

Site 1

10	5	10	5	12	13	9	12	11	8
10	13	13	14	12.5	13.5	13	15	7	9

Site 2

3	4	6	6	6	5	10	4	7	6
4	6	7	5	3	4	9	3	4	3

2 Outline the differences between the two samples.
3 Identify another graphical technique that could have been used to display this data.
4 Give reasons for your choice of technique.
5 The students also gathered information on the shape of the pebbles using Powers' roundness index (which gives a value to the shape of pebbles, see page 93). How could they have displayed this information on the same graph?

Foundation tier

1 Complete a dispersion graph for the information for pebble sizes in Site 1.
2 Describe the pattern shown by the data.
3 State another graphical technique that could be used to display this data.
4 Give reasons for your choice of technique.

Dispersion graphs are a sophisticated graphical technique which would rarely appear on foundation tier papers.

Review

By the end of this section you should be able to:

- construct dispersion graphs
- describe and explain the patterns that the graphs show
- use the graphs appropriately.

STRETCH AND CHALLENGE

Dispersion diagrams can be used to determine the range of data by plotting the inter-quartile range. Plot the median, upper and lower inter-quartile values and the inter-quartile range on the dispersion diagram for river data.

Proportional symbols

Learning objective – to study proportional symbols.

Learning outcomes
- To be able to construct proportional symbols.
- To be able to describe and explain the patterns that the proportional symbols show.
- To be able to suggest appropriate uses of these types of proportional symbols.

What is a proportional symbol?

These are symbols that are drawn in proportion to the size of the variable being represented. The symbol could be anything, for example a pictogram could be used, but is more usually circles or squares. This is a sophisticated data presentation technique and will usually only appear on higher tier papers.

How to draw proportional circles and squares on a map

- Find a base map of the area for the values you are displaying.
- Calculate the square root of each of the values. To work out the square root for Spain, put 14 into a calculator and press the square-root key. The answer should be rounded to one decimal place, so 3.7 is accurate enough.
- This answer will be used to construct your proportional symbol, either the radius of a circle or the side of a square.
- Study the square-root values to determine the range.
- Determine a scale that suits the range of square-root values. Remember the symbol must fit on the base map. For example, in Figure 18 a scale of 5 mm is used. Multiply each square-root value by 5 to determine the symbol size. Therefore, the proportional symbol for Spain has a radius of 18 mm.
- Locate the points on the map where you are going to draw your symbol.
- Draw the symbols and shade or colour them.
- Provide a key for the scale and put a title on your map.

Top ten overseas tourist destinations	Number of British tourist visits (rounded to the nearest million)
Spain	14.0
France	12.0
USA	4.0
Irish Republic	4.0
Italy	3.5
Germany	3.0
Portugal	2.0
Greece	2.0
Netherlands	2.0
Turkey	2.0

Remember on the exam paper you can be asked why a particular graphical technique is appropriate.

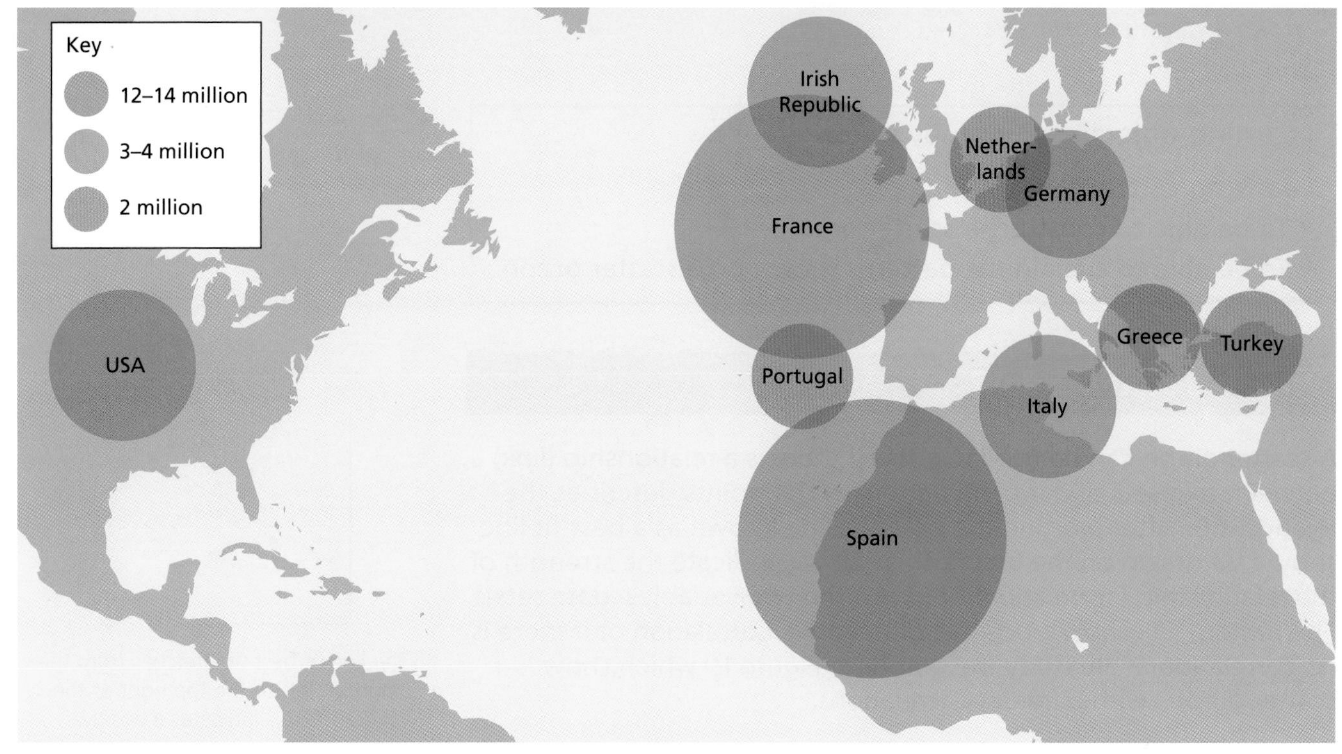

Figure 18 Proportional circles showing the number of British tourist visits to overseas tourist destinations

ACTIVITIES

1 Complete your own map of the top ten tourist destinations by following the 'how to' instructions.
2 Describe the pattern shown by the map.
3 State two advantages and two disadvantages of this technique.

STRETCH AND CHALLENGE

Complete a proportional circles map for the data in the table. You should draw pie charts for the information about water usage. The circles should be proportional to the gross domestic product in US dollars (GDP $) with the data within them.

Country	Domestic water usage (%)	Agriculture water usage (%)	Industry water usage (%)	GDP ($)
China	12	65	23	5,000
Australia	15	75	10	29,000
Japan	20	62	18	30,000
India	6	89	5	2,900
Russia	19	18	63	9,000
Uzbekistan	5	93	2	1,700
Malaysia	17	62	21	8,500
Sri Lanka	4	93	3	3,700
Afghanistan	3	96	1	700
USA	17	42	41	37,800

Review

By the end of this section you should be able to:

- construct proportional symbols
- describe and explain the patterns that proportional symbols show
- use the graphs appropriately.

Scatter graphs

Learning objective – to study scatter graphs.

Learning outcomes
- To be able to construct a scatter graph.
- To be able to explain the patterns shown on a scatter graph.

What is a scatter graph?

A scatter graph can be used to show if there is a relationship (link) between two sets of data. The pattern of the points describes the relationship. After plotting the points, a line known as a best-fit line should be drawn on the graph. This line will indicate the strength of the relationship (correlation) between the two variables (data sets). The pattern will show a positive or negative correlation or if there is no correlation at all. Study the graphs in Figure 19 which show scatter graphs with different correlations.

Positive correlation

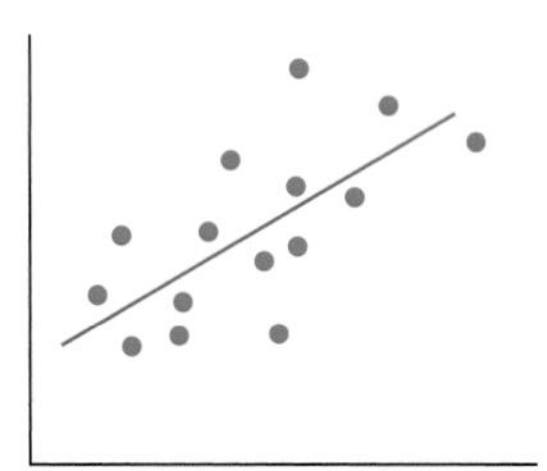

The line of best-fit stretches from the bottom left to the top right of the graph. This indicates a positive correlation; as one variable increases so does the other variable.

Negative correlation

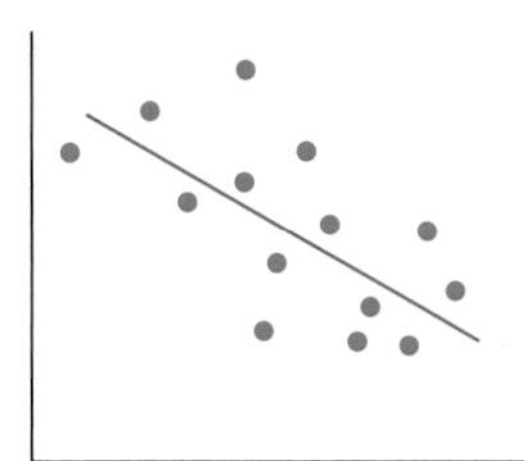

The line of best-fit stretches from the top left to the bottom right of the graph. This indicates a negative correlation; as one variable increases the other variable decreases.

No correlation

The points are distributed all over the graph. This shows that there is no relationship between the variables.

Figure 19 Scatter graphs with different correlations

How to draw a scatter graph to show if there is a graphical correlation between the width and depth of a river as it moves from its source (site 1) towards its mouth (site 10)

- Decide which is the independent variable and which is the dependent variable. For these two sets of data there is no independent or dependent variables. However, if you were plotting how depth changes with distance from the source, the distance from the source would be the independent variable and the depth would be the dependent variable.
- Decide on an appropriate scale on the *x*-axis for the width measurements. Remember, the scale should be spaced out evenly and allow for the highest value in the data set. In this case, ten squares on the graph paper equals 1 metre.
- Decide on an appropriate scale on the *y*-axis for the depth measurements. Remember, the scale should be spaced out evenly and allow for the highest value in the data set. In this case five squares on the graph paper equals 10 cm.
- Plot the measurements for each of the sites on to the graph, labelling each site with the correct number.
- Draw a line of best-fit. This is a straight line through the middle of the points that you have plotted.
- Compare the pattern with the standard patterns for the different types of correlations shown in Figure 19.
- What type of correlation have you plotted?
- Explain what this means.

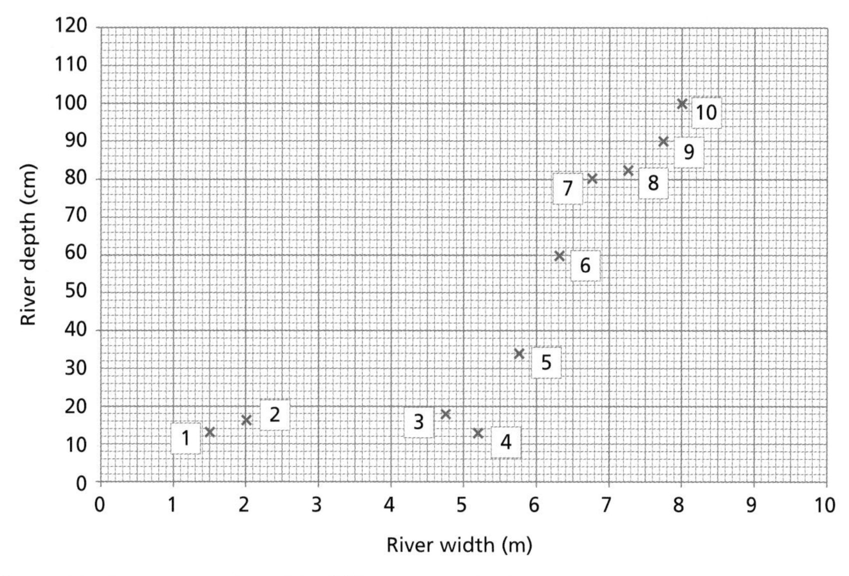

Figure 20 Scatter graphs with different correlations

Site	Average width (m)	Average depth (cm)
1	1.5	12
2	2.0	16
3	4.7	18
4	5.2	12
5	5.8	32
6	6.3	60
7	6.7	80
8	7.3	82
9	7.8	90
10	8.0	100

Topics	Scatter graphs can be used to show
Challenges for the Planet	Carbon dioxide emissions and GDP
Coastal Landscapes	Distance travelled along the beach and size of pebbles
River Landscapes	River depth and distance from source
Glaciated Landscapes	Deaths from avalanches and GDP
Tectonic Landscapes	Number of deaths from earthquakes and GDP
A Wasteful World	GDP against waste generated
A Watery World	GDP and industrial usage of water
Economic Change	GDP and percentage of population employed in tertiary industry
Farming and the Countryside	GDP and percentage of population employed in agriculture
Settlement Change	Distance from city centre and land values
Population Change	Life expectancy and literacy rate
A Moving World	Number of emigrants and number of people over 65 in the population
A Tourist's World	Average number of holiday nights and average number of holiday days per year

ACTIVITIES

Higher tier

1 Draw a scatter graph for the data below. Use the 'how to' box to help you.

Country	Domestic water usage (%)	GDP ($)
China	12	5,000
Australia	15	29,000
Japan	20	30,000
Thailand	9	7,400
Korea	14	17,700
India	6	2,900
Indonesia	8	3,200
Russia	19	9,000
Turkey	15	6,700
New Zealand	48	21,600
Uzbekistan	5	1,700
Malaysia	17	8,500
Sri Lanka	4	3,700
Algeria	25	5,900
Aghanistan	3	700
Sierra Leone	2	500
USA	17	37,800

2 Is there a correlation? What is its nature?

3 Points that are well away from the line of best-fit are known as residuals or anomalies. Are there any residuals or anomalies? If so, circle them on your graph.

4 Describe and give reasons for the pattern that is shown by the graph.

Foundation tier

1 Draw your own copy of the scatter graph for the river data on page 77. Use the 'how to' box to help you.

2 Complete the following sentences to explain the relationship shown by the graph.

The best-fit shows that there is a ...... correlation between the depth of the river and the width of the river. This means that as the width of the river ...... the depth of the river ...... Site ...... is an anomaly because the depth is ...... than site ...... which is closer to the source.

STRETCH AND CHALLENGE

Another way to test for a relationship between sets of data is to use a statistical technique such as Spearman's rank correlation coefficient. Test the statistical correlation between the data given in higher tier question 1 using Spearman's test. Information on how to complete this statistical technique can be found in the Stretch and Challenge section in Chapter 6.

Remember to always state the type of correlation and explain what it means.

Review

By the end of this section you should be able to:

- construct and interpret scatter graphs.

4 Geographical Enquiry and ICT Skills

In **Chapter 4 Geographical Enquiry and ICT Skills** you will learn how to study geographical enquiry and ICT skills.

Learning objective – to study geographical enquiry and ICT skills.

Learning outcomes

- To understand the skills required to carry out a geographical enquiry and how those skills can be examined.
- To understand the ICT skills which can be used to aid and enhance geographical enquiries.

Geographical enquiry and ICT skills will be examined when you complete your controlled assessment. However, there could also be questions relating to them in the exam paper that tests geographical skills. You will need to be able to show your understanding of the processes involved in carrying out a geographical enquiry. You will not be expected to have access to a computer in the exam, but may have to show your understanding of the principles and use of ICT in geographical enquiries. Questions could be asked on how to extract information from the internet or how to use databases such as the census.

Geographical enquiry skills

The following are the geographical enquiry skills you will be expected to understand as part of your controlled assessment and they could be examined on the geographical skills paper.

Identify, analyse and evaluate geographical questions, hypotheses and issues

All geography students should be able to identify geographical questions. For example, if you look at a particular location, what are the questions you should be asking yourself:

- What is the landscape like?
- What are the features that stand out?
- Where is this place – grid references?
- What is the area like around it?
- Why is it like it is?
- What is happening to certain variables?

From asking geographical questions you should be able to form hypotheses on which a piece of work could be focused. A hypothesis is a statement that can be tested. In an exam question, you might be asked to formulate hypotheses from a sample.

A geographical issue is a debatable point. For example, should a wind farm be located in a particular place? Questions such as the following could then be asked:

- What will be the impact on the environment?
- What will be the impact on local people?
- Are the climatic conditions correct?

Hypotheses could be set up and tested, or questions could be set and answered, but it is unlikely that a clear answer will be determined because so many different viewpoints are involved. Issues usually require candidates to make judgements from inconclusive data or evidence.

Whether you are studying a geographical question, hypothesis or issue you should be able to explain why you are studying it and the results you expect to find.

You could be given this task question and asked to develop hypotheses to help answer it.

How do channel characteristics vary along a river?

Sample hypotheses for the rivers question:

- The velocity of the river decreases as it moves from its source to its mouth.
- The width of the river increases as it moves from its source to its mouth.
- The depth of the river increases as it moves from its source to its mouth.
- The gradient of the river channel increases as it moves from its source to its mouth.
- The wetted perimeter of the river increases as it moves from its source to its mouth.

The next stage would be to explain why each of the hypotheses is relevant.

Establish sequences of enquiry

All geographers should be able to establish a sequence of enquiry. These are sometimes called road maps. What it means is planning out how you will complete your investigation. You must establish what you need to do to answer your task question and the techniques you will need to complete to derive the data you require. You also need to plan when you will be doing each stage of the enquiry and set yourself deadlines. When you write up your study it should be in the same sequence with each section in a different chapter. There should be good linkage between the chapters which shows the sequence of the enquiry.

You could be asked to plan the sequence of a geographical enquiry.

A brief sample road map for a river study:

Planning stage:	location and hypotheses
Data collection:	when should I go on my field trip?
	what techniques should I use?
Data presentation:	what techniques shall I use?
Analysis:	explaining my results
Conclusions:	answering my questions
Evaluation:	what was the value of my study?
	how successful were my techniques?

Extract and interpret information from a range of sources including field observations, maps, drawings, photographs, diagrams and tables and secondary sources

It is important that you are able to extract information from many different sources; this will be important for all parts of your course. In the skills section you could be asked to extract information from photographs and sketches in relation to OS maps. In other questions you could be asked to interpret information from different sources including fieldwork.

Ask yourself these questions:

- What can you see? In other words, what are the overall patterns or main features?
- Are there specific groupings of features or information?
- Are there any anomalies or features which are particularly different?

You could be given some river data as in the table below:

Site	Width (m)	Average depth (cm)	Channel gradient (degrees)	Wetted perimeter (cm)
1	1.5	13	7	171
2	4.7	18	6	540
3	5.3	12	5	590
4	8.5	92	1	10.8

A question might ask you to extract and interpret data from a table.

The following answer extracts the data:
'The river gets wider as it flows towards its mouth. It was 1.5 metres at the first site and 8.5 metres at site 4.'

The following answer interprets the data:
'As the river gets wider it also gets deeper; an anomaly is site 2 where it suddenly becomes deeper. This may be because of the particular point in the river where we took the measurement which was on the outside of a meander bend.'

Describe, analyse and interpret evidence

During your controlled assessment you will accumulate data which you will have to make sense of in the analysis and conclusions section of your work. In the skills paper you might be given some data or a graph. The question could ask you to analyse it. Follow these points:

- Describe what you see.
- Ensure that you have included important facts such as dates or numbers or even names.
- Give reasons for the patterns that you see in the data.
- Try to link your reasons to other data that you have collected.

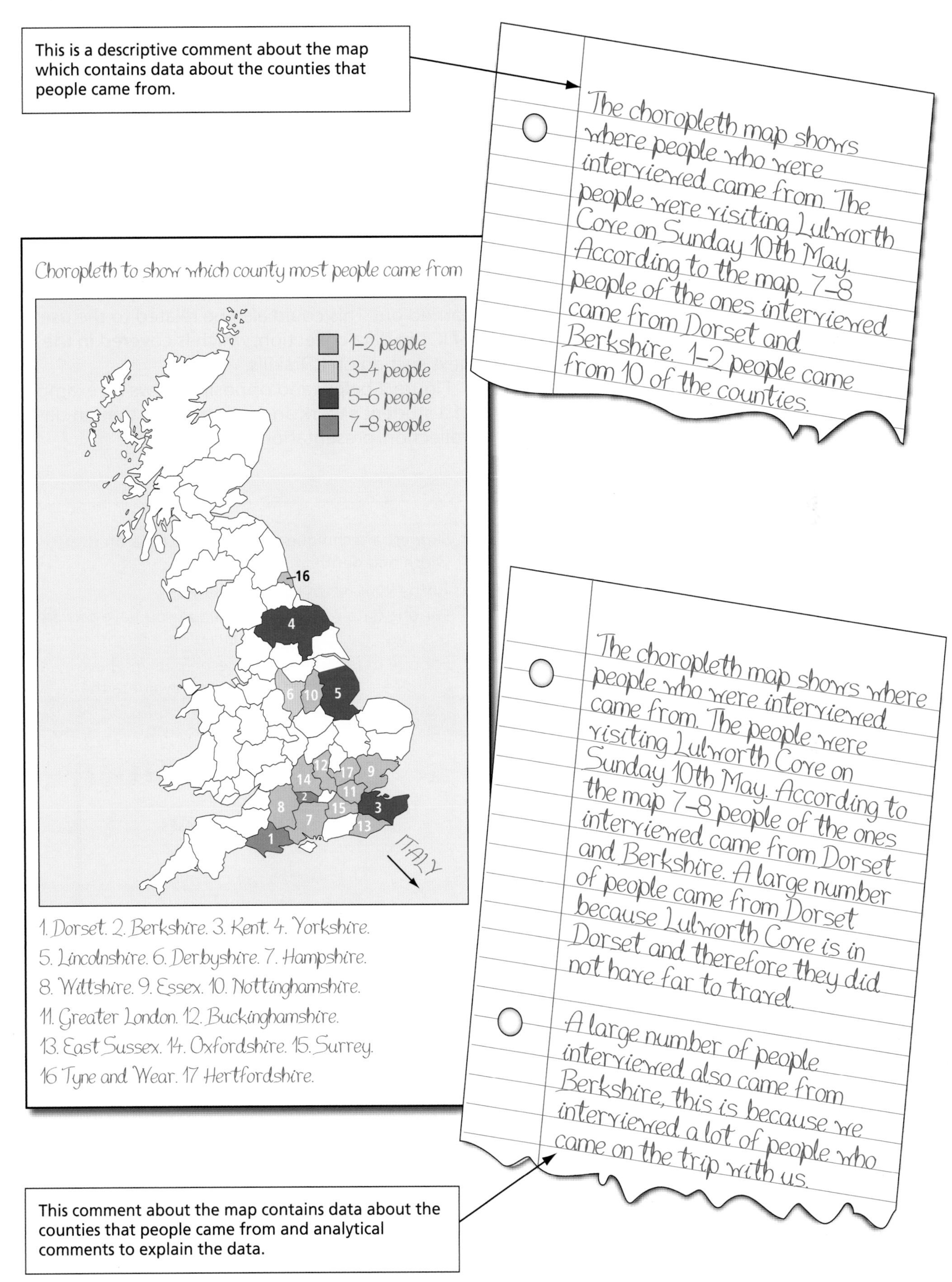

Figure 1 Example of student's work. A question might ask you to describe and interpret a choropleth map

Draw and justify conclusions from evidence

The conclusion is when you return to your original hypotheses or questions and answer them using the evidence you have provided in the study. The more evidence you use to back up your findings the more plausible they will be. On the skills paper you could be given some findings and asked to answer a hypothesis based on those findings.

Evaluate methods of data collection, presentation and analysis of evidence

In your controlled assessment you will be required to evaluate or state the value of the data collection techniques, the data presentation techniques and the study as a whole, in other words, the evidence you have managed to find out.

In the skills paper you may be asked to do an evaluation of a particular type of data collection or presentation technique. You may also be asked to evaluate the findings of a particular study. When completing evaluations it is always a good idea to focus on how things could be improved or the problems you had and how you might solve them.

Questions on data collection technique could be linked to your controlled assessment by asking you to describe and explain a technique you have carried out. This could also be related to the use of ICT in data collection, which is covered in the next section on ICT skills.

Figure 2 below and opposite shows an example of a student's work answering a question on data collection, presentation and analysis.

ACTIVITIES

1. Devise hypotheses for the following issue. Should a tidal barrage be built in the Severn Estuary?
2. Evaluate the usefulness or justify your answer to question 1.
3. What would be the sequence of enquiry for a task which was investigating the presence of longshore drift on a beach?
4. Suggest a technique to display the data on river width and depth.
5. Justify your answer to question 4.
6. For one data collection technique you have carried out, describe, explain and evaluate the method.
7. For one data presentation technique you have completed, describe, explain and evaluate the method.

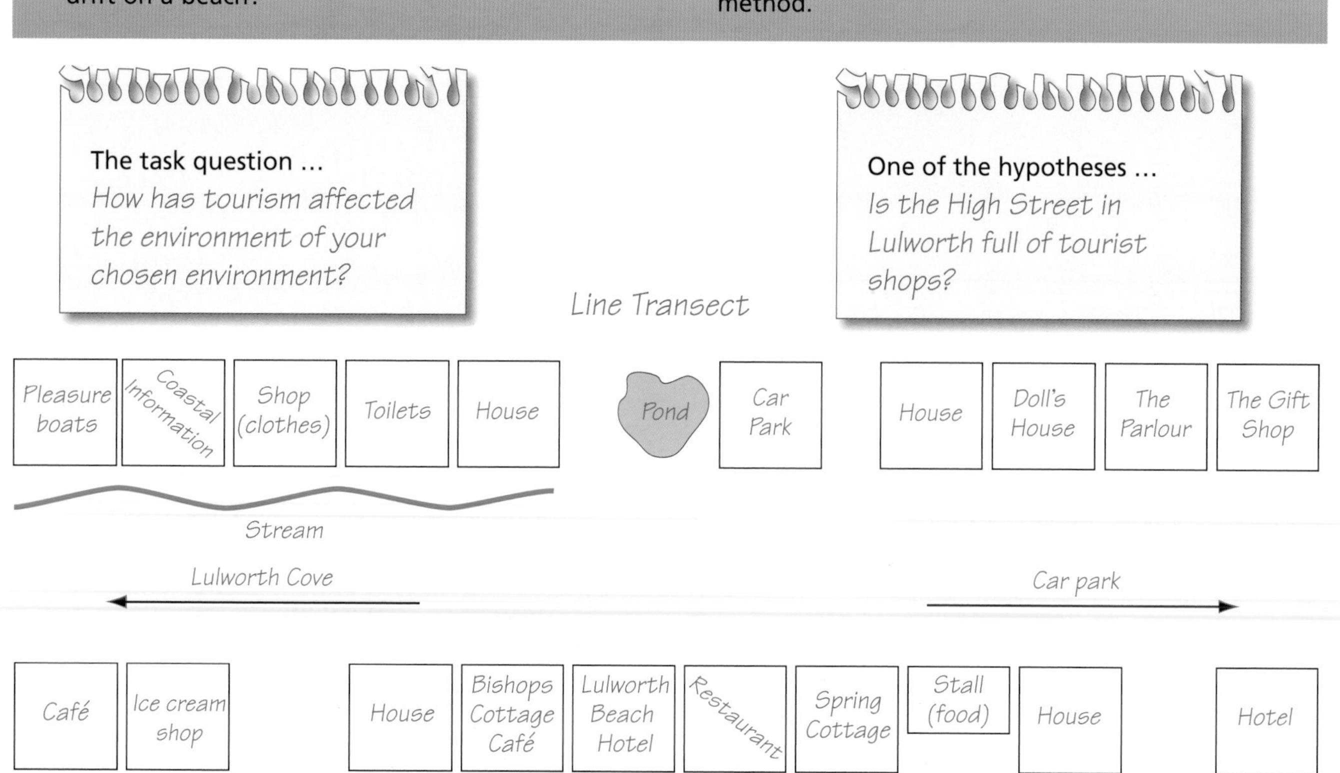

Figure 2 **(a)** Example of a student's work. A question might ask you to draw conclusions from evidence and evaluate methods of data collection, presentation and analysis

The data collection method …

The data was collected by walking down the High Street noting down what was on either side. A category chart with a key was used which helped to identify the different shops. The key meant that we had room on our chart to fit everything in. Each person did their own transect because they could have interpreted the High Street in different ways.

Evaluation of the method …

This is a good way of collecting the data because it would be very reliable as it was primary data collection. However, I did have problems giving all of the shops a category; I asked other members of my group for help.

Here the method of data collection is explained and evaluated.

The data presentation method …

I drew out each of the buildings on the High Street and wrote in the box what it was. I marked clearly which end of the street was near the car park and which end was by the beach.

Evaluation of the method …

This method could have been improved if I had drawn everything to scale. The road would then have been the correct length and the buildings would have represented their actual size. I could then have assessed their importance to the area. I should also have taken photographs of the buildings to help explain their use.

Here the method of data presentation is described and evaluated.

The analysis of your findings …

The line transect shows that there are many tourist related shops in the High Street. They provide food and souvenirs for the tourists, there were no shops that provided for the local community.

A simple analytical comment which uses the evidence on offer to put forward a viewpoint.

A simple concluding comment. In normal circumstances more than one technique should be used to draw concluding comments.

Concluding the hypothesis …

From the evidence of the line transect the High Street is full of tourist shops. Therefore the tourists are having an impact on the character of Lulworth Cove.

Figure 2 (b) Example of a student's work. A question might ask you to draw conclusions from evidence and evaluate methods of data collection, presentation and analysis

ICT skills

The following are ICT skills which will be tested through your use of them when completing your controlled assessment:

- Collect and annotate photographs and satellite images.
- Use the internet, for example to investigate case studies of volcanic eruptions and floods.
- Extract information from video and television programmes.
- Research and present investigative work.

Exam questions will focus on how ICT can aid geographical studies and provide stimulus material where appropriate. You will not be expected to remember the results of your controlled assessment study. However, it might be useful as an example in answering some of the questions.

Use spreadsheets and data-handling software in the field

ICT can be used to plan the data collection with the use of word processors to produce questionnaires and provide spreadsheets to enable data to be easily collected in the field. Data loggers can be used in the field to collect vast amounts of data such as temperature, speed of river flow or simple traffic counts. The information collected can then easily be transferred on to spreadsheets and displayed.

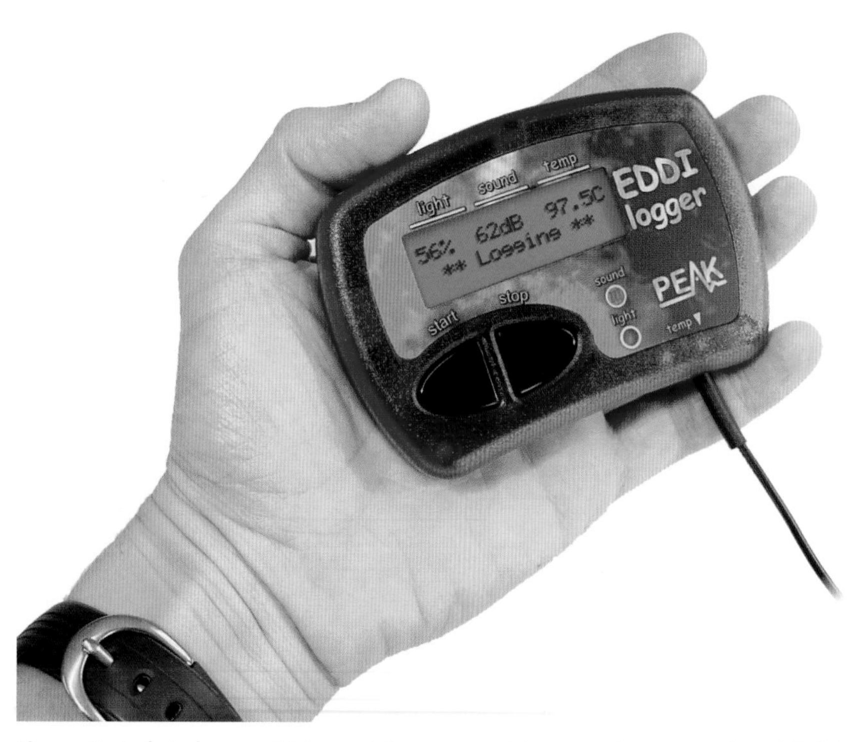

Figure 3 A data logger. This model can record temperature, sound and light intensity

Carry out data presentation and analysis techniques

Information collected in the field electronically on data loggers can easily be presented using appropriate software packages. ICT can also be used to present data, for example a spreadsheet can be used to provide graphical techniques from questionnaires. There are also programs which will produce beach profiles and kite diagrams. On the skills exam paper, questions could focus on the advantages and disadvantages of using ICT in the field to collect data and in the classroom to display data. Questions could also be set using stimulus material provided on these areas.

Use databases such as census and population data

These databases have a wealth of information that can be used in your controlled assessment, but parts of them could also be used in examinations. Census data could be given and you could be asked to interpret it or to compare one ward within an urban area with another. You could also be asked to complete choropleth maps with information from the census.

ACTIVITIES

1 How does ICT help with data collection in the field?
2 How does ICT help with data presentation?
3 a Trace a copy of the incomplete map of Swindon's wards from Figure 4 on page 88.
  b Use the data in the table to map the percentage of white people who live in the selected wards of Swindon.
  c Use tracing paper to overlay the information on the percentage of people of working age.
4 Describe the age structure of the different wards.
5 Comment on the age structure of the wards in relation to their geographical position.

STRETCH AND CHALLENGE

Suggest reasons for the patterns shown by the data for the selected wards of Swindon.

Review

By the end of this section you should be able to:

- understand the skills required to carry out a geographical enquiry and how those skills can be examined
- understand the ICT skills which can be used to aid and enhance geographical enquiries.

Electoral ward	Male (%)	Female (%)	Christian religion (%)	White ethnicity (%)	Aged 0–15 years (%)	Aged 16–64 years (%)	Aged over 65 years (%)
Abbey Meads	53	47	66	92	26	71.5	2.5
Blunsdon	51	49	77	99	17	62	21
Eastcott	53	47	61	92	15	73	12
Central	55	45	60	90	10	75	15
Shaw & Nine Elms	50	50	66	93	27	69	4
Wroughton & Chiseldon	50	50	77	98	21	61	18
St Margarets	51	49	79	97	20	63	17
Freshbrook	50	50	68	94	24	67	9
Gorse Hill & Pinehurst	51	49	64	94	22	65	13
Ridgeway villages	51	49	77	98	23	64	13

Figure 4 Map of some of Swindon's electoral wards

5 Geographical Information System (GIS) Skills

In **Chapter 5 Geographical Information System (GIS) Skills** you will learn how to study Geographical Information System (GIS) skills.

The people indicate where newspapers are delivered to each day.

This layer shows houses. The central building is a newsagent.

This layer shows the roads of the area.

Base map showing physical features.

Learning objective – to study Geographical Information System (GIS) skills.

Learning outcomes

- To understand how geographical information is captured and represented.
- To know how to use web mapping tools.

What is GIS?

GIS touches every part of our daily life without many people realising it. The term is daunting but the way it works isn't. GIS is a way of using maps digitally to make our lives easier. It allows a large amount of information to be seen by layering one set of information on to another. The base information is a map of the area being studied. It can be used on PCs and even on mobile phones. One of the most common uses of GIS is Global Positioning Systems (GPS) used by thousands of car drivers and hill walkers.

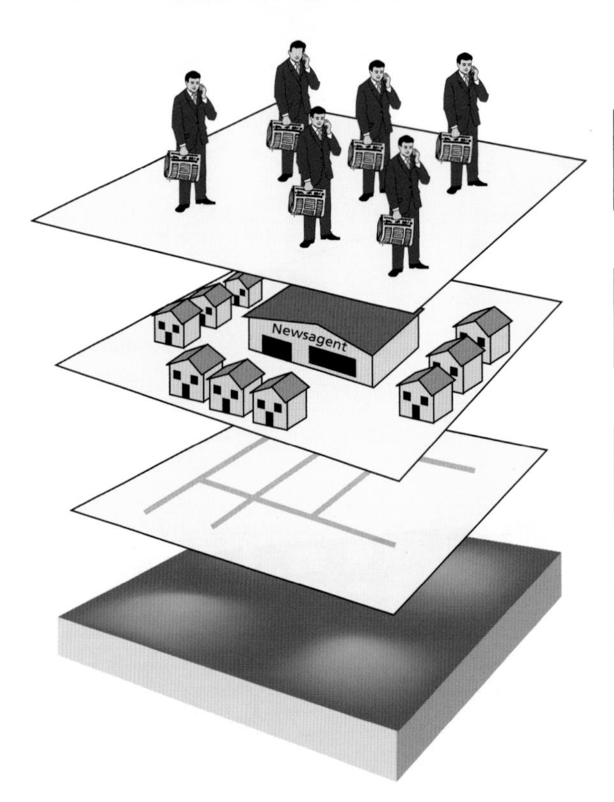

The people indicate where newspapers are delivered to each day.

This layer shows houses. The central building is a newsagent.

This layer shows the roads of the area.

Base map showing physical features.

Figure 1 A GIS map. A GIS map has a number of layers that can be placed on each other to build a picture of an area

The advantages and disadvantages of GIS

Advantages	Disadvantages
A lot of information can be seen on one map	The information can become difficult to see if too much is put on to one map
Information can be linked together easily to form patterns that can be analysed	A computer or other ICT equipment is needed as the maps are digital
GIS is available on iPhones and other types of mobile phone	The equipment is expensive to buy and keep up to date
GIS is of great benefit to many public services such as the police and utilities	A certain amount of training is needed to use the more sophisticated systems
GPS has made travelling between places much easier	

Some uses of GIS

How have ambulance stations used GIS to improve their response times?

- A call comes into ambulance central control. The call-handler uses GIS to quickly find where the patient is located.
- The next map layer tells central control where all the other ambulance crews are at the moment.
- The next layer on the map is traffic congestion: central control may decide to send a crew who are further away but will get there quicker due to traffic congestion in certain areas.

GIS has been used to research where accidents have happened in the past and therefore where would be the best place to have ambulances located at certain times of the day. This has also improved response times of ambulances across the country.

How have public utility companies used GIS to improve their service to customers?

- The location of public utilities such as telephone lines and water pipes are all mapped.
- The information is shared on a central, secure website.
- The different utilities have access rights to the website. They can look to see what other utilities are in the areas where they wish to work.
- Maintenance work is also logged on the website. Different companies are encouraged to plan so that routine maintenance can be carried out at the same time.

Exam questions are likely to be about the processes and principles of GIS, not about using GIS on a computer. You could be asked what GIS is, who might use GIS, what layering means or even the advantages or disadvantages of GIS.

ACTIVITIES

1 What do the letters GIS stand for?
2 How does GIS work?
3 What is meant by the term 'layering'?
4 State four organisations or people who use GIS.
5 You order a pizza from a pizza delivery company. How could GIS be used to deliver your pizza?
6 A supermarket chain wants to locate a new store in an area. The following layers were used to work out the best place for the store. Put the layers into the order that you think they would have been used. Justify your order:
   - population density
   - land for sale
   - transport routes
   - competitors
   - socio-economic groupings of residents.
7 In Unit 2 you will study either waste or water. The following website has a lot of information on the USA which can be layered: http://nationalatlas.gov/natlas/Natlasstart.asp.
   a If you are studying 'A Watery World':
      - Go to the website.
      - Click on Basic maps: Cities and Towns
      - Click on Water, followed by Dams.
      - The key to the map can be found by pressing the tab at the top, Map key.

      Describe the distribution of dams in the USA.
   b If you are studying 'A Wasteful World':
      - Go to the website.
      - Click on Basic maps: Cities and Towns.
      - Click on People, followed by Energy Consumption.
      - The key to the map can be found by pressing the tab at the top, Map key.

      Describe the residential energy consumption in 2001 in the USA.

GIS will be tested in an exam by using stimulus material. You will need to know about what GIS is and who benefits from its use.

STRETCH AND CHALLENGE

Why not try some of the GIS missions on the Ordnance Survey website? http://mapzone.ordnancesurvey.co.uk/mapzone/giszone.html

Review

By the end of this section you should be able to:
- understand how geographical information is captured and represented
- know how to use web mapping tools
- be aware of the types of examination questions you could be asked.

6 Stretch and Challenge Information

This chapter provides further information and details so that you can extend your skills by learning more about some of the techniques mentioned in the book.

Powers' roundness index

Maurice Powers published a paper in 1953 describing a technique of measuring pebbles. This technique gives a visual representation against which the shape of pebbles can be compared. The chart, now called the Powers' roundness index, is split into six categories, each denoting a degree of roundness from very angular to well rounded. One disadvantage is that one person's opinion on the roundness or angularity of a pebble may differ from another's.

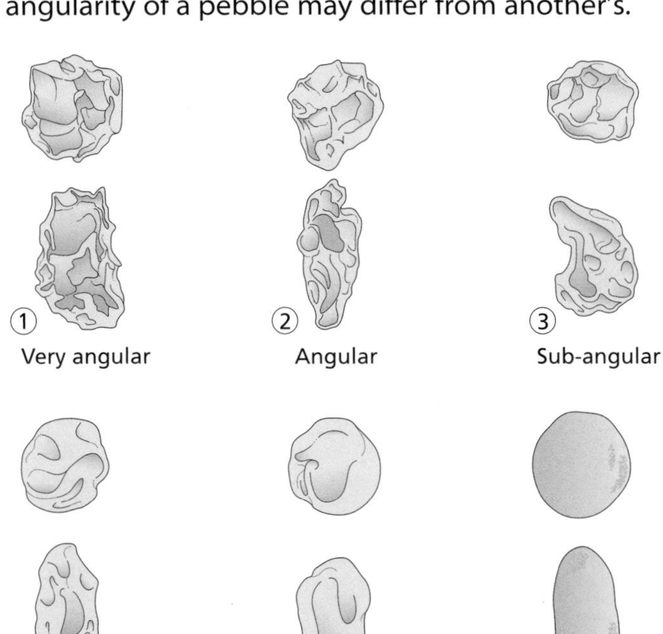

Figure 1 Powers' roundness index

How to work out the inter-quartile range of a set of data

The inter-quartile range (IQR) shows the spread of a set of data, but is more accurate than using the range because it doesn't include the extremities. The range is the difference between the highest and lowest value in the data set. The IQR indicates the spread of the middle 50 per cent of the data set as it omits the top and bottom 25 per cent. It gives a better idea of how the data is spread around the median value.

The IQR is calculated by putting the data into rank order or plotting it on a graph. The values are then divided into four equal groups or quartiles, where n is the number of values.

The upper quartile (UQ) value is the value that occurs at the following position in the data set:

$$\frac{(n+1)}{4}.$$

The lower quartile (LQ) value is the value that occurs at the following position in the data set:

$$\frac{3(n+1)}{4}.$$

The difference between these two values is known as the IQR.

Worked example

Consider the rainfall data for a city in Israel shown in Figure 2. This data could either be ranked or drawn as a graph to work out the IQR.

Year	Average rainfall (mm)	Rank
2008	200	3
2006	140	1
2004	452	10
2002	190	2
2000	290	6
1998	442	9
1996	347	7
1994	560	11
1992	390	8
1990	209	4
1988	235	5

Figure 2 Rainfall data for a city in Israel

To work out the IQR for the rainfall data:

$UQ = 11 + 1 = \frac{12}{4} = 3$

Third position = 200

$LQ = 3 \times 12 = \frac{36}{4} = 9$

Ninth position = 442

Therefore the IQR is 242 mm and the median is the middle value (see Figure 3).

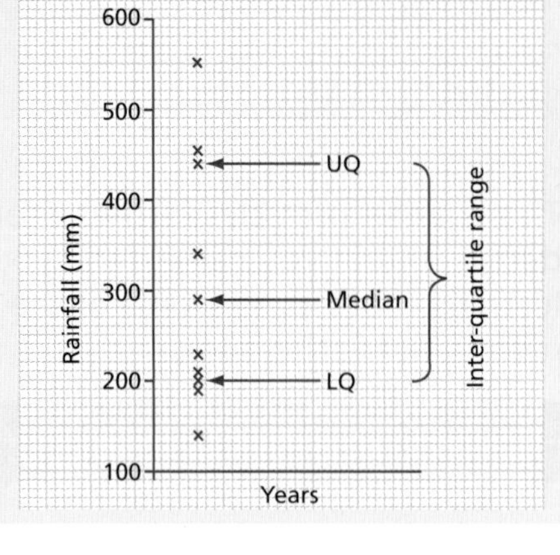

Figure 3 Graph of the rainfall data

Spearman's rank correlation coefficient

Spearman's rank correlation coefficient is a technique that is used to see if there is a link between sets of data.

Here is a worked example to find out if there is a correlation between settlement size and number of services in a settlement.

Step 1. The data should be put into a table, as shown, with the smallest number in the first set of data appearing first in column A, with its corresponding value in column C.

A Settlement population	B Rank	C Number of services	D Rank	E Differences between ranks (d)	F d^2
190	12	4	11	1	1
330	11	3	12	1	1
1,550	10	11	10	0	0
2,660	9	19	9	0	0
5,632	8	39	8	0	0
8,890	7	41	7	0	0
9,950	6	87	4	2	4
10,500	5	43	6	1	1
10,900	4	92	3	1	1
11,890	3	75	5	2	4
14,330	2	109	2	0	0
16,789	1	144	1	0	0

Step 5. The differences should then be totalled. In this example this is a result of 12. So $\Sigma d^2 = 12$. You will need this in the next part of the calculation.

Step 4. The difference between the rankings, column E, is then squared and added to column F.

Step 2. Both sets of data should be given a rank order, with the largest observation given 1, as is the case for 16,789. This settlement also has the largest number of services, so it also has a rank of 1. Ranking should be placed in columns B and D.

Step 3. Work out the difference between the ranks of each variable and add to column E.

Step 6. Calculate the coefficient (r_s) using the formula:

$$r_s = 1 - \left(\frac{6\Sigma d^2}{n^3 - n} \right)$$

where Σ is the symbol for summation, d is the difference in rank of the values of each matched pair and n is the number of pairs.

Remember that $d^2 = 12$ from the above calculation, and $n = 12$ (the number of pairs of data). Therefore:

$$1 - \left(\frac{6 \times 12}{12^3 - 12}\right) = 1 - \left(\frac{72}{1728 - 12}\right) = 1 - \left(\frac{72}{1716}\right) = 1 - 0.04$$

$$= 0.96$$

The result can be interpretated from the scale:

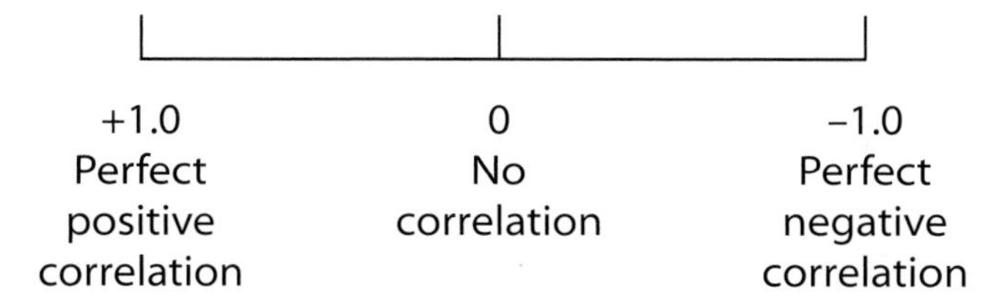

In the above example a result of +0.96 indicates that there is a strong positive correlation between settlement size and the number of services at these sites.

Further development of the technique

You should now determine if the correlation has happened by chance or not. To do this you must decide on the rejection level (α), this is, if you like, how confident you want to be about your findings. Therefore, if you wish to be 95 per cent certain, your rejection level is calculated as follows:

$$\alpha = \frac{100 - 95}{100}$$

$$= 0.05$$

Calculate the formula for t:

$$t = r_s\sqrt{\left(\frac{n - 2}{1 - r_s^2}\right)}$$

where r is Spearman's rank correlation coefficient (= 0.96), n is the number of pairs (= 12) and df is the degrees of freedom (this means how confident you are about your results). Therefore, with a rejection level of 0.05 or 95 per cent:

$$t = r_s\sqrt{\left(\frac{n - 2}{1 - r_s^2}\right)} = 0.96\sqrt{\left(\frac{12 - 2}{1 - 0.96^2}\right)} = 10.73 \text{ (the critical value)}$$

You should now work out the degrees of freedom:

df = $n - 2$

where n = number of pairs = 12.
Therefore

df = $(n - 2) = (12 - 2) = 10$.

Look this up in the *t*-table (Figure 4) using the degrees of freedom of 10 and a 0.05 rejection level. Therefore, the critical value is 10 and the *t*-value on the *t*-table is 2.23.

So, the *t*-value is less than the critical value which means there is a significant correlation between settlement size and the number of services found in them.

If the critical value is less than your *t*-value then the correlation is significant at the level chosen. If the critical value is more than your *t*-value then you cannot be certain that the correlation did not occur by chance. This could be because:

- your sample was too small to permit you to prove a correlation
- the relationship is not a good one to have chosen.

Degrees of freedom	Rejection level probabilities				
	$p = 0.1$	$p = 0.05$	$p = 0.02$	$p = 0.01$	$p = 0.001$
1	6.31	12.71	31.82	63.66	636.62
2	2.92	4.30	6.97	9.93	31.60
3	2.35	3.18	4.54	5.84	12.94
4	2.13	2.78	3.75	4.60	8.61
5	2.02	2.57	3.37	4.03	6.86
6	1.94	2.45	3.14	3.71	5.96
7	1.90	2.37	3.00	3.50	5.41
8	1.86	2.31	2.90	3.36	5.04
9	1.83	2.26	2.82	3.25	4.78
10	1.81	2.23	2.76	3.17	4.59
11	1.80	2.20	2.75	3.11	4.44
12	1.78	2.18	2.68	3.06	4.32
13	1.77	2.16	2.65	3.01	4.22
14	1.76	2.15	2.62	2.98	4.14
15	1.75	2.13	2.60	2.95	4.07
16	1.75	2.12	2.58	2.92	4.02
17	1.74	2.11	2.57	2.90	3.97
18	1.73	2.10	2.55	2.88	3.92
19	1.73	2.09	2.54	2.86	3.88
20	1.73	2.09	2.53	2.85	3.85
21	1.72	2.08	2.52	2.83	3.82
22	1.72	2.07	2.51	2.82	3.79
23	1.71	2.07	2.50	2.81	3.77
24	1.71	2.06	2.49	2.80	3.75
25	1.71	2.06	2.49	2.79	3.73
26	1.71	2.06	2.48	2.78	3.71
27	1.70	2.05	2.47	2.77	3.69
28	1.70	2.05	2.47	2.76	3.67
29	1.70	2.05	2.46	2.76	3.66
30	1.70	2.04	2.46	2.75	3.65
40	1.68	2.02	2.42	2.70	3.55
60	1.67	2.00	2.00	2.66	3.46

Figure 4 A *t*-table

ACTIVITY

Work out if there is a correlation between the distance from a groyne on a beach and the size of shingle on the beach.

Distance from groyne (m)	Size of shingle (cm)
50	7.6
100	7.9
160	8.2
300	6.8
420	7.4
480	6.8
550	5.8
670	7.9
850	6.8
950	4.7
1050	5.6
1150	4.2

Remember if you have a joint rank (that is when two numbers in the table have the same value) you should add them together and then divide by the number there are, missing out the next numbers until you have omitted how many you add together.

Answers

Chapter 1 Basic Skills

Label and annotate diagrams, graphs and sketch maps (page 4)

1

2 Simple sketch with the following marked on it:
- a The built-up areas of Alnmouth and Warkworth.
- b The Rivers Aln and Coquet.
- c The coastline and all coastal features.
- d Any tourist information features in the area.

Drawing and labelling sketches (page 7)

1

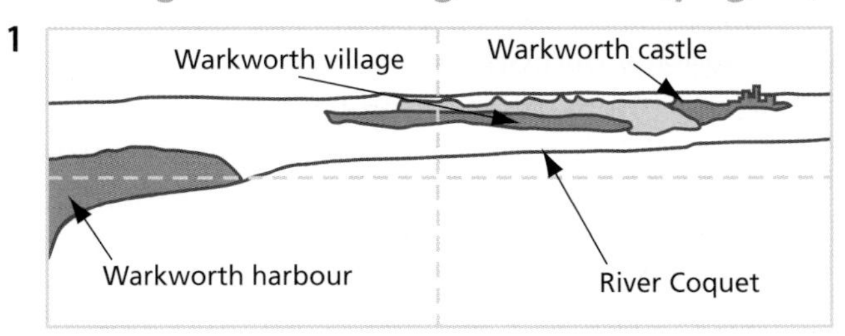

2

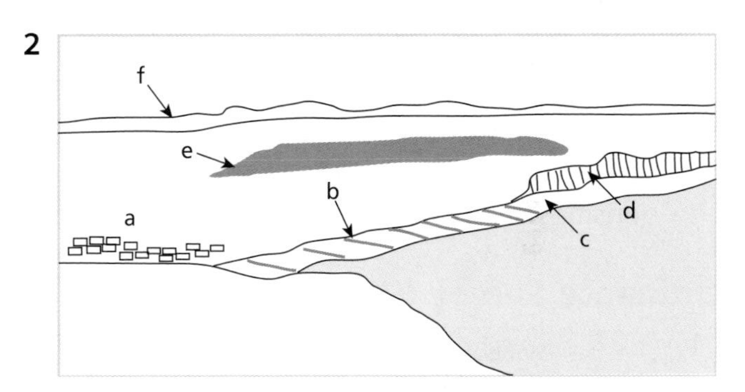

Interpretation of aerial, oblique and satellite photographs (page 11)

1

Oblique aerial photograph	Vertical aerial photograph	Satellite photograph
Shows a larger area but less defined. Shows outline of coastline, hills and lowland areas. Position of settlements, wooded areas and fields.	Shows a smaller area but more detail. Does not show if the area is flat or hilly. Coastline shown clearly. Shows areas which are fields and areas that are built up. Shows actual houses, roads/ street patterns.	Shows a large area of the coastline. Can distinguish beaches and landforms to some extent. Can identify built-up areas, field patterns and wooded areas.

2 Human features: castle, defensive site, village in the centre of the meander bend, large fields around the village. Physical features: flat land, no sign of animals so probably arable (crop-farming), wide bend (meander) of the river, coastline with sandy beach in the background which is possibly used by tourists in the summer.

Chapter 2 Cartographic skills

Atlas maps (page 14)

1 a The tropical rain forests are found either side of the Equator within the Tropics. The continent with the largest rainforest area is South America in Brazil. Europe does not have any rainforests.
b The higher land is found to the north and west of Great Britain. It is more evenly spread in Ireland. The land above 1000 m is mainly found in Wales and Scotland with none found in the south and east of Britain. The highest land in England is in the south-west and north of the country.
c Names of mountainous areas, cities.
d The highest population densities are found on the eastern coastal areas and the river valleys in the west, for example, Yangtze. The lowest densities are in areas such as the Gobi Desert.

Ordnance Survey (OS) maps (page 18)

1

Symbol	Feature
TH	Town Hall
	Viewpoint
	Tunnel
	Electricity transmission line
	Non-coniferous woodland
CH	Club house

2

Symbol	Key	Four-figure reference	Six-figure grid reference
	Place of worship with tower	5683	567833
P	Post office	5885	587857
△	Triangulation pillar	5884	582847
	Golf course or links	5785	571853
□ Home Fm	Home Farm	5583	551839
	Bus or coach station	5685	568858
□ Sch	School	5784	571841

3

Feature	Symbol
Flat rock	

4 National Trust (always open).
5 Coniferous.
6 PH – Public House, P – Post Office, Place of worship (church) with tower.
7

Tourist feature	Grid reference
Information	187133
Caravan site	190119
Camp site	190121
Parking	187142
(Alnwick) Castle	189138
Museum	196122

8 North-west.
9 South.
10 South-west.

Straight line and winding distances (page 19)

Wensleydale

1 13.4 cm, 6.7 km.
2 12.6 cm, 6.3 km.
3 The river.
4 0.4 km.
5 10.8 cm, 5.4 km.

Swanage

6 14.6 cm, 7.3 km.
7 The A road distance is 10.2 cm, 5.1 km. Therefore the B road is longer.
8 2.2 km.

Cross-sections (page 22)

1 a–c See cross-section below. **2, 3** See cross-section below.

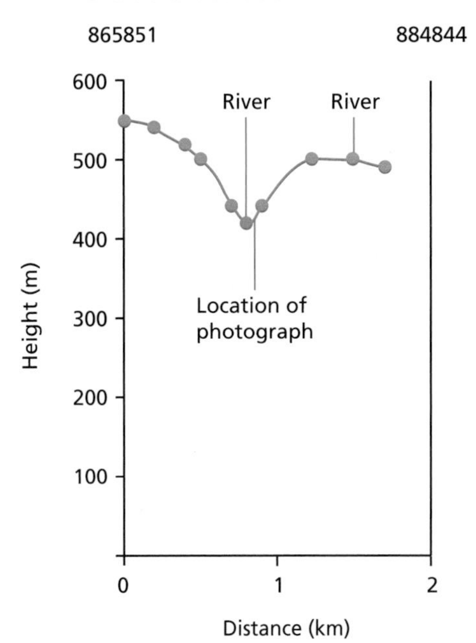

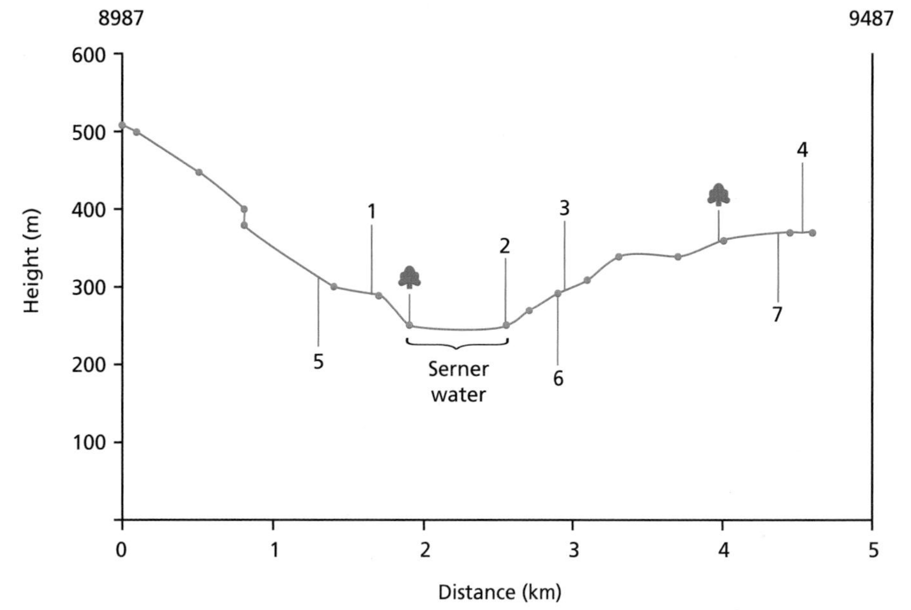

d South-west.
e Shape of the hills, small valley floor.

The distribution and pattern of human and physical features (page 25)

Looe map

1 The woodland is found along the river valleys. West Looe river. There is no woodland on the hilltops. There is no woodland above 130 m. There are some small patches of woodland, for example 2458.

2 The railway line follows the main river valley on the map. The East Looe River. The A387 follows this river for about 3 km then takes the route around the top of a hill to the west. Other routes mainly follow the river valleys or pass between the hills.

Swanage map

3 The main settlement, Swanage, is on the coast. It is between the ranges of hills, one to the north and one to the south. There are small settlements on the slopes of the southern hills such as Worth Matravers (9777). The other larger settlements in the area are Corfe Castle which has grown close to a gap in the northern hills that allows access to the area. There is also Harman's Cross which is in the valley halfway between the other two settlements.

4 Straight boundaries, roadways can be seen through the woodland which allows management. Local knowledge or map evidence provides screen for the oil field to the north.

5 Roads on either side of the steep hill. The main A351 goes through the main valley. Few roads to the south of the map, steep hills and lack of settlement.

Site, situation and shape of settlements (page 27)

1 Where a number of rivers meet (at the confluence of four or five rivers). A bridging point on the West Looe river. On both sides of the river approximately 40–60 m above sea level.

2 Herodsfoot is a nucleated settlement with housing around the place where the rivers meet. Duloe is a linear settlement which has houses built either side of the road B3254.

3 Swanage is situated on the coast. The A351 links it to Corfe Castle. There are other smaller roads leading north out of the town. The main area of the town is on flat land next to the coast although there is new development on the southern slopes.

Human activity from map evidence (page 28)

1 Post Office, public house, place of worship with a spire, mile stone, stone circle.

2

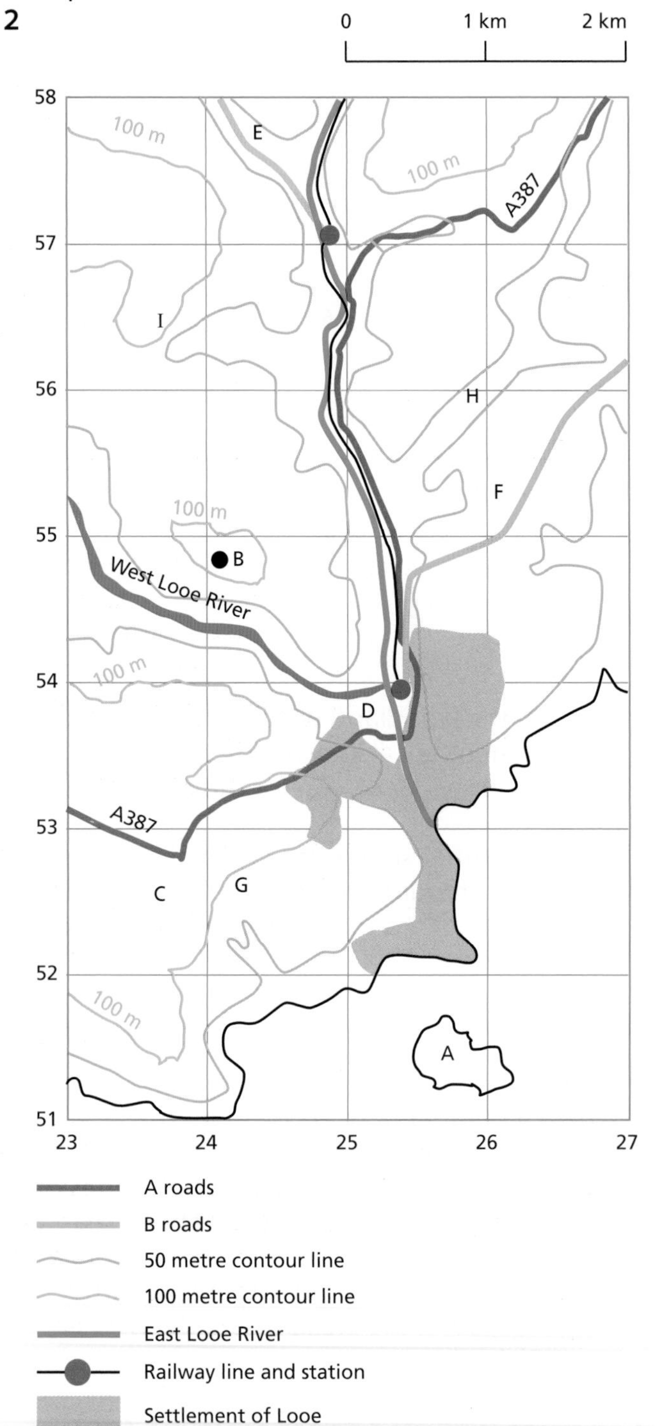

3

Place	Feature on map
Island A	St. George's or Looe
Spot height B	115 m
Tourist attraction C	Caravan and camping site
Tourist attraction D	Parking
Road number E	B3254
Road number F	B3253
Settlement G	Port Looe
Type of woodland H	Non-coniferous
Farm I	Puffiland Farm

Use maps in association with photographs, sketches and directions (page 31)

1 a North-west.

b

Letter on photograph	Grid reference	Feature
V	Six figure = 254535	Number of road = A387
W	Four figure = 2454	Name of woodland = Trenant wood
X	Six figure = 257530	Identify symbol = Beacon
Y	Four figure = 2554	Name of river = East Looe River
Z	Six figure = 251537	Identify symbol = Parking

c Many possible answers; cars parked, the actual buildings, wooded areas around housing to west of river which are not on the map.

d Many possible answers, such as museum, road numbers, names of villages.

2 They leave Corfe Castle on the B3351 towards Studland. After they have travelled for 2 km, they pass Brenscombe Farm on their right. They arrive in Studland and park the car in the car park closest to the Post Office at grid reference 038826. The family walk in an easterly direction to Old Harry in grid square 0582. They leave Studland and travel to Swanage. In the village of Ulwell they pass a camp site and a caravan site. At the end of the day they return to their hotel. They leave Swanage on the A315 to Corfe Castle.

3 Take the road that is less than 4 m wide by the church without a spire or tower out of Herodsfoot heading towards Duloe. They will go up a steep hill with a gradient of 1 in 5 or steeper. Turn right at the first T-junction. They

will then pass through Coombe Farm going down a hill between 1 in 7 and 1 in 5. Continue straight through Duloe, passing onto the B3254. This road meets the A387 at Sandplace, continue south to Looe. Follow the road over the bridge; there is a car park to the right. Turn right out of the car park and continue on the A387. The A387 ends in grid square 2051, take a left turn. This road leads to Polperro and is a dead end. Be careful you don't drive into the sea!

Sample Examination Questions

Cambridge map (page 32): Higher tier

1 a 4657. (1 mark)
  b Railway station. (1 mark)
  c 472572. (1 mark)
  d A1134. (1 mark)
  e The housing is in a grid-iron pattern. This means that the housing is in blocks. The streets are in a line and meet each other at right angles. (3 marks)

2 a Cul-de-sacs, curves, not in a straight line compared to comments in 1e. Comparative comment not required, implicit comparison is satisfactory. (3 marks)
  b The housing is on cheaper land at the edge of the city. (1 mark)

3

	On map not on photographs B or C	On photograph B or C not on map
Feature 1	Churches	Fields
Feature 2	Road numbers	Types of buildings

(2 marks)

4 The 'park and ride' sites are all on the outskirts of Cambridge. They are close to major roads that lead into the city. There is a 'park and ride' site in grid square 4259. This is close to a motorway junction. (3 marks)

Cambridge map: Foundation tier

1 a 4657. (1 mark)
  b Railway station. (1 mark)
  c 472572. (1 mark)
  d A1134. (1 mark)
  e The housing is in a grid-iron pattern. The houses are in blocks. The streets are in a line. (3 marks)

2 a Cul-de-sacs, curves, not in a straight line compared to comments in 1e. Comparative comment not required, implicit comparison is satisfactory. (2 marks)
  b The housing is on cheaper land at the edge of the city. (1 mark)

3

	On map not on photographs B or C	On photograph B or C not on map
Feature 1	Churches	Fields
Feature 2	Road numbers	Types of buildings

(2 marks)

4 Outskirts, major, 4259, motorway. (4 marks)

Swanage map (page 34): Higher tier

1 a B3351. (1 mark)
  b 035822. (1 mark)
  c National Trust. (1 mark)
  d Camp and caravan site. (1 mark)
  e 5. (1 mark)

2 Harman's Cross is a linear village. Most of the houses are either side of the main A351 road. The village of Worth Matravers has a more nucleated shape with the houses being around a junction. (3 marks)

3 Appropriate examples: the A351 from Swanage to Corfe Castle follows the valley and avoids the high ground but stays above actual floor close to the river; the B3351 and parallel minor road avoid Brenscome Hill and Nine Barrow Down; and Swanage to Studland road cuts through gap between Ballard Down and Nine Barrow Down. (2 marks)

4 Sketch (1 mark), accurately shading the town of Swanage (1 mark), beach and groynes (1 mark) and Swanage Bay (1 mark), Peveril Point (1 mark), the pier (1 mark).

Swanage map: Foundation tier

1 a B3351. (1 mark)
  b 035822. (1 mark)
  c National Trust. (1 mark)
  d Camp and caravan site. (1 mark)
  e 5. (1 mark)

2 Linear, A351, nucleated, junction. (4 marks)

3 Sketch (1 mark), accurately shading the town of Swanage (1 mark), beach (1 mark), groynes (1 mark), Peveril Point (1 mark), Swanage Bay (1 mark).

Warkworth map (page 36): Higher tier

1 a South.
  b River Coquet. (2 marks)

2 See sketch map for answers. (5 marks)
  e The river valley is at a height of 10 m. The river is in its lower course near the sea. The river has many (meander) bends. A railway

line crosses the river in grid square 2212. In grid square 2311 the river is crossed by the A1068. The river flows to the south of Lesbury. There is a small coniferous woodland to the south of the river in grid square 2211. (4 marks)

3 Leave Hermitage Farm, turn left onto the secondary road. After approximately 1 km arrive at a junction with the A1068. Turn left onto the A1068 towards Alnmouth, drive for approximately 5 km. Go straight across the roundabout in grid square 2311 continuing on the A1068. After approximately 5 km you will then reach the town of Alnwick.

Continue through the town on the B6341, parking in the car park close to the castle. (4 marks)

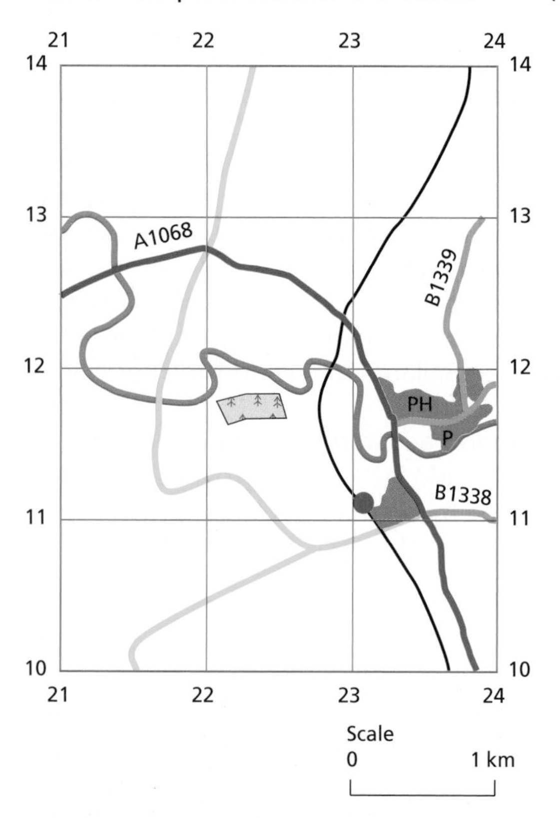

Warkworth map: Foundation tier

1 a i South (1 mark)
b iv Coquet. (1 mark)

2 See sketch map. (4 marks)
c 2212, A1068, south, south. (4 marks)

3 Leave Hermitage Farm, turn left on to the secondary road.

After approximately 1 km arrive at a junction with the A1068.

Turn left onto the A1068 towards Alnmouth, drive for approximately 5 km.

Go straight across the roundabout in grid square 2311, continuing on the A1068.

After approximately 5 km you will then reach the town of Alnwick.

Continue through the town on the B6341, parking in the car park close to the castle. (5 marks)

Looe map (page 38): Higher tier

1 S = St. George's or Looe Island
T = Flat rock. (2 marks)

2 It is a wet point site. It has a spring in the village which would have supplied water to the villagers. It is above the floodplain of the river. It is on the junction of a number of secondary roads. (4 marks)

3 One mark for each label and one mark for drawing the cross-section. (5 marks)

4 a Museum, IRB station, information centre, place of worship with a tower. (3 marks)
b Boats, benches, promenade. (1 mark)

Looe map: Foundation tier

1 a Looe Island
b Flat rock. (2 marks)

2 Wet, spring, drinking, above, secondary. (5 marks)

3 One mark for each label and one mark for drawing the cross-section. (5 marks)

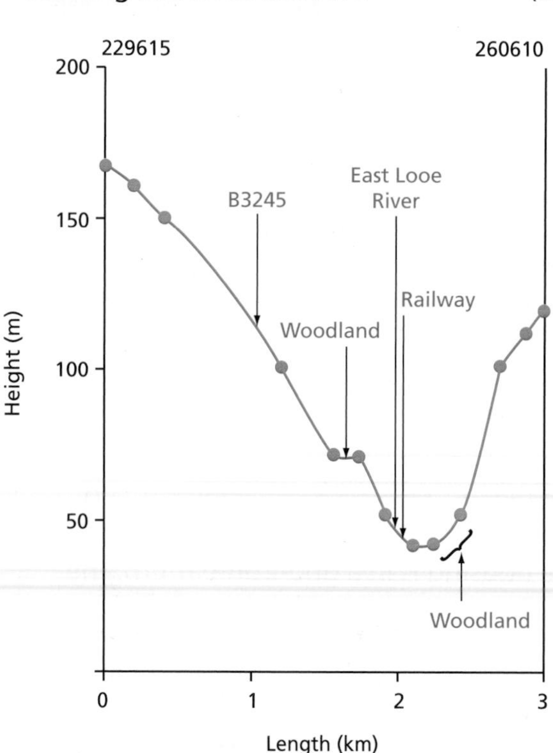

4 a Museum, IRB station, information centre, place of worship with a tower. (3 marks)
  b Boats, benches, promenade. (1 mark)

Wensleydale map (page 40): Higher tier

1 a River Ure. (1 mark)
  b West to east. (1 mark)

2

Service	Hawes	Bainbridge
Public house	✓	✓
Place of worship with a spire	✓	
Visitor centre	✓	
Post Office	✓	

(4 marks)

3 The settlements are in the valleys because that is where the main road is (the A684). The settlements are not on the hills because they are too steep. This is shown by the contour lines being close together. The settlements are found in the river valleys to have easy access to water for drinking. (4 marks)

4 The River Ure bends. It has a wide flat valley floor about 0.5 km. The river flows from west to east. There is a caravan site to the south of the river. It is approximately 220 m above sea level. Cragdale water has a number of smaller bends than the Ure which has one big one in the section we are looking at. It has steep sides and very little flat land in the valley bottom. The river flows from south-east to north-west. (5 marks)

Wensleydale map: Foundation tier

1 a River Ure. (1 mark)
  b West to east. (1 mark)

2

Service	Hawes	Bainbridge
Public house	✓	✓
Place of worship with a spire	✓	
Post Office	✓	

(3 marks)

3 Any three tourist features from the map key: visitor centre, craft centre, etc. (3 marks)

4 The settlements are in the valleys because that is where the main road is (the A684). The settlements are not on the hills because they are too steep. This is shown by the contour lines being close together. The settlements are found in the river valleys, such as the River Ure, to have easy access to water for drinking. (2 marks)

5 Small, west to east, south, steep, south-east to north-west. (5 marks)

Chapter 3 Graphical Skills

Different types of bar graph (page 53)

Higher tier

1 Population pyramid accurately drawn.

2 There are large numbers of people in the 20–29 age group: 15 per cent males and 12 per cent females. There are very few people over the age of 69: only 13.5 per cent.

3 It is easy to see the numbers in each age group and to get a quick visual impression of the numbers. Numbers are continuous.

4 Flow lines.

5 Pedestrian counts are flows of people, therefore flow lines are the appropriate technique.

6 All of the countries have a winter maximum. The amount that they have in summer is very low, the highest being 80 mm in Rome. The lowest summer total is in Tripoli with 1 mm. The highest winter total is Tel Aviv, the lowest being 125 mm in Tripoli. Tel Aviv has the greatest difference between summer and winter precipitation, Rome the smallest.

Foundation tier

1 Population pyramid accurately completed.

2 Males, 3 per cent, females, 1 per cent.

3 It clearly shows if any of the age groups has more of the population than the others. The data is continuous.

4 All of the countries have a winter maximum. The amount that they have in summer is very low, the highest being 80 mm in Rome. The lowest summer total is in Tripoli with 1 mm. The highest winter total is Tel Aviv, the lowest being 125 mm in Tripoli. Tel Aviv has the greatest difference between summer and winter precipitation, Rome the smallest.

Line graphs, compound line graphs, flow line graphs and isolines (page 59)

Higher tier

1 Advantages: shows at a glance the peaks and flows in the traffic, comparisons between times are easy to make, easy to construct, continuous data.

  Disadvantages: can be a problem if the number of vehicles varies greatly between the times, would be better if displayed as a flow because it doesn't show direction, which a flow would.

2 **a** Flow line map.
  b Shows direction as well as numbers of vehicles.
  c Many more people went into Lulworth on Sunday than any other flow. The peak flow was 136 vehicles between 2p.m. and 2.30p.m. The lowest flow is out of Lulworth on Friday p.m. between 3p.m. and 3.30p.m., when ten vehicles left Lulworth.

3 Arrows should go along the streets where the people walked. There is no scale on the key. The arrows are different lengths. It is not clear from the location of the arrows where the counts were actually completed. The number of pedestrians should not be written on the arrows, it should be determined from the key.

4 Map redrawn taking the points in question 3 into consideration.

5 Histogram.

6 Compound line graph drawn as in the 'how to' box.

Foundation tier

1 Advantages: shows at a glance the peaks and flows in the traffic, comparisons between times are easy to make, easy to construct, continuous data.

  Disadvantages: can be a problem if the number of vehicles varies greatly, would be better if displayed as a flow because it doesn't show direction, which a flow would.

2 **a** Flow line map.
  b Shows direction as well as numbers of vehicles.
  c Many more people went into Lulworth on Sunday than any other flow. The peak flow was 136 vehicles between 2p.m. and 2.30p.m. The lowest flow is out of Lulworth on Friday p.m. between 3p.m. and 3.30p.m., when ten vehicles left Lulworth.

3 Arrows should go along the streets where the people walked. There is no scale on the key. The arrows are different lengths. It is not clear from the location of the arrows where the counts were actually completed. The number of pedestrians should not be written on the arrows, it should be determined from the key.

4 Map redrawn taking the points in question 3 into consideration.

5 Histogram.

Pie diagrams (page 61)

Higher tier

1 Pie diagram as in Figure 11.

2 The number employed in primary industry has declined from about 8 per cent in 1991 to about 2 per cent in 2006. The number in secondary industry declined between 1991 and 2001. It was then fairly stable between 2001 and 2006. Tertiary industry has increased over the whole period from about 63 per cent in 1991 to 85 per cent in 2006.

3 Compound bar chart.

4 Gives a visual impact of change over time better than pies. Easier to construct because no working out of angles.

Foundation tier

1 Pie diagram as in Figure 11.

2 8 per cent, 27 per cent, decreased, increased.

3 Lines could be drawn for each type of industry, the peaks and troughs can be easily seen.

Pictograms (page 63)

Higher tier

1 Could be pictures of people but better if pictures are of different vehicles.

2 Bar chart.

3 Pictures of vehicles show clearly what has been counted. Numbers of vehicles can be seen easily by the key.

4 Data is more accurately presented. Much easier to construct.

Foundation tier

1 Pictogram drawn.

2 Bar chart drawn.

3 Pictures of vehicles show clearly what is being displayed. Numbers of vehicles can be seen easily by the key.

Rose/ray diagrams (page 65)

Higher tier

1 Ray diagram drawn.
2 The greatest number of tourists went to France with 11 million visitors. The least number of tourists went to the Netherlands and Turkey with only 2 million. The only destination in the top ten that is not in Europe is the USA with 4 million tourist visits.
3 Shows the direction as well as the numbers of people.
4 A histogram would show the countries and numbers clearly.
5 The techniques will both show the data clearly but the ray diagram will also show the direction of the destinations from the UK. The length of the rays showing the number of tourists would be much more effective than drawing bars.

Foundation tier

1 Ray diagram completed.
2 The greatest number of tourists went to France with 11 million visitors. The least number of tourists went to the Netherlands and Turkey with only 2 million. The only destination in the top ten that is not in Europe is the USA with 4 million tourist visits.
3 Continuous data. A histogram would clearly show the differences.
4 The techniques will both show the data clearly but the ray diagram will also show the direction of the destinations from the UK. The length of the rays showing the number of tourists would be much more effective than drawing bars.

Triangular graphs (page 67)

Higher tier

1 Triangular graph constructed.
2 The country with the highest number of people employed in the primary sector is Ethiopia with 84 per cent. The country with the lowest number of people employed in primary industry is the UK with 2 per cent. The country that is spread most evenly between the sectors is Mexico. The country with the largest number employed in secondary is Japan. France has the second highest tertiary sector with 66 per cent employed.
3 They are very difficult to construct.

   They are able to show three sets of data on one graph.

   They can be confusing if you are unsure which way to read the lines.

   They have to be constructed for data which adds up to 100 per cent.

   It is very easy to see the patterns they portray.

   Anomalies are easy to spot.

   Base triangles have to be provided because they are very difficult to construct.

   They can save a lot of time and effort because a lot of information is on one graph.
4 Nine.

Foundation tier

1 Triangular graph constructed.
2 Ethiopia, 2 per cent; Mexico, Japan, 66 per cent.
3 Advantages: they are able to show three sets of data on one graph. It is very easy to see the patterns they portray. Anomalies are easy to spot. They can save a lot of time and effort because a lot of information is on one graph.

   Disadvantages: they are very difficult to construct. They can be confusing if you are unsure which way to read the lines. They have to be constructed for data which adds up to 100 per cent. Base triangles have to be provided because they are very difficult to construct.

Topological diagrams (page 69)

Higher tier

1 Answer will be different for each student.
2 They condense the information allowing more information to be shown in a small space. They allow inaccuracies.
3 a At Waterloo get on the Northern line. Go through six stations, get off at the seventh station which will be Euston. Details of stations in between may be given.
   b Go to Embankment on the District Line. Change to the Northern line, go north to Tottenham Court Road. Change to the Central line, go east to Liverpool Street. Details of stations in between may be given.
4 You have no idea of the exact direction on the ground that you are travelling. You have no idea of the distance you have travelled, as no scale is given.

Foundation tier

1 Answer will be different for each student.
2 **a** Go east on the Hammersmith and City line.
  b Go south on the Bakerloo line.
3 One disadvantage is that you have no idea of the distance you have travelled. One advantage is they condense the information allowing more to be shown in a small space. The London Underground map enlarges the middle and shrinks the outside.

Choropleth maps (page 71)

Higher tier

1 Choropleth drawn.
2 Advantages: shows immediately at a glance the pattern made by the data, visual representation is easy to complete.
  Disadvantages: only shows a spread of data not an amount, the ranges of data used can make the map ineffective in showing the pattern.
3 Bars on a map of the world.
4 The bars show immediately by their height the amount; with a choropleth the key would need to be referred to. The choropleth does not show a specific amount but a range of data.
5 Continuous data.

Foundation tier

1 Choropleth drawn.
2 Most tourists to France with 81.9 million tourists visiting there. The tenth most popular world tourist destination is Mexico with 21.4 million tourists. This is the least to expect. There should be other comments about countries within these figures.
3 Advantages: shows immediately at a glance the pattern made by the data, visual representation is easy to complete.
4 Flow line map completed.
5 Continuous data.

Dispersion graphs (page 73)

Higher tier

1 Dispersion graph completed.
2 Pebbles are larger at site 1. The largest pebble at site 2 is still as large as the middle range of pebbles at site 1. The smallest pebble is 3 mm at site 2 but site 1 has a pebble only 2 mm bigger. Therefore there is little difference between the sizes.
3 Scatter graph.
4 It would also show the dispersion of the pebbles.
5 The dots on the graph could have been a certain shape. Therefore drawing a pictogram of the shape of the pebble on the graph where the size was located.

Foundation tier

1 Dispersion graph completed.
2 Pebbles are larger at site 1. The largest pebble at site 2 is still as large as the middle range of pebbles at site 1. The smallest pebble is 3 mm at site 2 but site 1 has a pebble only 2 mm bigger. Therefore there is little difference between the sizes.
3 Scatter graph.
4 It would also show the dispersion of the pebbles.

Proportional symbols (page 75)

1 Completed map.
2 The greatest number of tourists went to France with 11 million visitors. The least number of tourists went to Portugal, Greece, the Netherlands and Turkey with only 2 million. The only destination in the top ten that is not in Europe is the USA with 4 million tourist visits.
3 Advantages: shows clearly the differences due to the different sizes of the circles, shows location so the spread of the data can be seen at a glance.
  Disadvantages: difficult to construct. Data is shown in a range not an actual value.

Scatter graphs (page 78)

Higher tier

1 Scatter graph drawn.
2 There is a positive correlation.
3 New Zealand, Algeria.
4 If a country has piped water to houses, more water will be used. It is a percentage therefore that will alternate with the other uses which are agriculture and industry. If the country is a dry one the agricultural percentage will be higher.

Foundation tier

1 Scatter graph drawn.
2 Positive, increases, increases, 4, less, 3.

Chapter 4 Geographical Enquiry and ICT Skills

Geographical enquiry skills (page 84)

1 Hypotheses such as:
The barrage would have a detrimental effect on fish stocks in the area.
The barrage would have a negative visual impact on the area.
2 If the barrage is built it may impede fish and therefore stocks of fish will decrease; it will impact on the area and could discourage tourists as many may see it as ugly.
3 Devise techniques to test for the presence of longshore drift, carry out the field work, collate results, present the data, analyse, conclude and evaluate your work.
4 Scatter graph.
5 It can show a relationship between two pieces of data.
6 Answers will vary depending upon technique chosen.
7 Answers will vary depending upon technique chosen.

ICT skills (page 87)

1 ICT can make the recording of the data more accurate. It can also make it easier to record because of the use of digital recorders. It can also look more professional if, for example, a typed questionnaire is used to interview people.
2 It can be an efficient way of displaying data using spreadsheets such as Excel to produce graphs.
3 Graph produced as stated.
4 The highest percentage of working age is in Swindon Central. The highest percentage of over 65 is in Blunsdon. The highest percentage of under 16 is in Abbey Meads.
5 The 'younger' population is in the centre of Swindon. The older population is on the edge of Swindon.

Chapter 5 Geographical Information Systems (GIS) Skills (page 92)

1 Geographical Information System.
2 It works by allowing different features to have their own layer on a map which can be put together in any order. This allows you to look at specific features or build a picture of an area to your requirements.
3 The ability that GIS gives you to put different types of information on a map by overlaying on a computer.
4 Ambulance service, supermarket chains, public utility companies, police.
5 A satellite navigation system is GIS; this would be used by the delivery person to find your address.
6 Land for sale, population density, socio-economic groupings of residents, transport routes, competitors.
The land has to be first because otherwise cannot build the store. The others could be in a different order but if no-one lives there then there is no point in putting a shop there. Comments such as this.
7 There are many different answers concerning the distribution of either dams or energy consumption. They should refer to dense and sparse areas and mention cities and areas of the USA. Refer to website: http://nationalatlas.gov/natlas/Natlasstart.asp.

Index